Tim Skern

Coffee House Notes on Virology

Tim Skern

Coffee House Notes on Virology

2nd edition

facultas.wuv

Bibliografische Information der Deutschen Nationalbibliothek

Die Deutsche Nationalbibliothek verzeichnet diese Publikation in der Deutschen Nationalbibliografie;
detaillierte bibliografische Daten sind im Internet über
http://dnb.d-nb.de abrufbar.

2nd edition 2009

Illustrations: Helmut Dolznig
Printed in Austria
ISBN 978-3-7089-0532-7

Preface to the second edition

Virology has not stood still in the four years since this book was written. Highly pathogenic H5N1 influenza A virus has continued to pose a global threat whereas an outbreak of hand, foot and mouth disease in China threatened the Olympic games in 2008. Vaccines against human papillomaviruses were approved in 2007. However, an outbreak of measles in Austria in 2008 demonstrated once again that levels of vaccination in Western societies are insufficiently robust to prevent outbreaks of disease. These events have however taken a back seat in the wake of the current influenza A virus pandemic. The world apprehensively awaits the 2009 winter influenza virus season in the Northern hemisphere.

The second edition contains three new questions. One examines why certain viruses have a polymerase in the particle whereas others do not. A second question looks into influenza virus attachment to avian and human cells and the origins of the current influenza A virus pandemic. The third question reflects the anniversary of 2009, highlighting how viral replication supports the work of Charles Darwin.

Several errors in the first edition have been corrected. I am grateful to the students who pointed them out to me. I also thank Tanja Kostić for proofreading the new questions.

Tim Skern, Vienna, October 2009

Preface

Teaching virology is an exciting and stimulating challenge. Daily media reports document the importance of viruses to the global health of all organisms. The diversity of viruses and their strategies for multiplication amaze and fascinate student, teacher and researcher alike. Research into virology is proceeding at a breathtaking speed; the results have profound implications for society in areas such as medicine, politics, ethics and economics. At the same time, these very attributes which make virology so fascinating also combine to make the subject such a challenge to teach. What and in how much detail should the student know about viral diversity? What are the fundamental principles which can be used across virology? Which themes are central to all topics within virology and related subjects?

"Coffeehouse notes on virology" is an attempt to provide the student with the basic principles and facts required for understanding viruses. These basic principles and facts can be found in the form of answers to 75 examination questions which I have used at the Universities of Graz and Vienna. To keep to the essentials, I wrote out each answer by hand with a limit of one A3 piece of paper and did not use books or other sources whilst writing. I wrote many of the original answers in coffeehouses, hotels, trains and occasionally at home. To ensure that the answers were comprehensible to the intended readership, five students then checked both the science and the language of the answers. Three of them (studying microbiology, medicine and molecular biology) had previously taken a course in virology whereas two (studying biology and nutritional science) had not. None of the students are native English speakers. The book should thus be easily accessible to all science and medicine students worldwide, even those with no prior knowledge of virology. I hope too that the book will stimulate interested readers to explore further the world of virology for themselves. As a teacher, "Coffeehouse notes on Virology" is the book that I would like to have had when I started in 2002 to prepare an entirely new course in molecular virology!

This book would not have been possible without the efforts and support of many students, friends and colleagues. I am grateful to you all. Special thanks to Friederike Turnowsky who first suggested that I teach a virology course at the University of Graz as well as to the five students, Tanja Kostić, Christiane Mair, Florian Mair, Angelika Mühlebner and Margarita Smidt, for reading the texts and making such invaluable suggestions. Tanja Kostić also typed in all the texts, made the layout consistent and was wonderfully honest in telling me whether an answer was clear or not. Hannes Tkadletz initially suggested how to layout the answers for typing. I am indebted to Sigrid Neulinger at Facultas Universitätsverlag for the relaxed and enthusiastic way she guided this project into print. Thanks to Helmut Dolznig for making my hand-drawn sketches into legible and printable graphics. Christian Mandl of the Medical University of Vienna was kind enough to proofread the book and made invaluable suggestions and comments. Wendy Burgers of the University of Cape Town carefully checked chapters 3 and 6; Dave Westaway of the University of Toronto made suggestions for the two answers on prions in his own inimitable style. I am grateful to all three of you. Any errors which remain in the book are entirely my responsibility.

Finally, I would like to mention the inspiration provided by Lauren Sompayrac (author of the book "How pathogenic viruses work"), Edwin Cameron (author of the book "Witness to AIDS") and Monwabisi Maqogi. Monwabisi is a passionate teacher on the subject of HIV and counsellor to AIDS sufferers in the township of Khayelitsha (near Cape Town) as well as in the Eastern Cape.

All my earnings from this book will be passed on to support Monwabisi's work.

Tim Skern, Vienna, October 2005

To the reader

The answers to the 78 questions contain what to me are the most important concepts, principles and facts in virology. Many questions contain new information whilst some present information found earlier in the book in a different way. Some questions are very similar but answered using different viruses. Important concepts and recurring ideas are stressed in blue.

If you are just starting out in virology, you may want to read the questions first before trying to answer them yourself. If you have some prior knowledge, try to answer the questions yourself first! Don´t worry if your answers are different to mine; many questions can be answered using different virus families or virus-host interactions as examples. The most important thing is that you can structure your knowledge of virology to see common features in the face of viral diversity.

I hope you find the book useful and that you enjoy this introduction to this fascinating subject!

Contents

2 RNA viruses

3 Retroviruses

4 DNA viruses

5 Viruses and cancer

6 Viruses and the immune system

7 Prevention and control of viral diseases

8 Evolution, emerging viruses and other agents

9 Emerging Questions

1 Fundamentals

1.1 What are viruses? Which cellular processes can not be performed by viruses? Outline the key steps found in viral life-cycles.

What are viruses?

- Viruses are obligate intracellular parasites (obligate = have no choice)
- Viruses are made up of nucleic acid, protein and sometimes lipid
  - e.g. common cold virus (human rhinovirus) comprises just protein and RNA
  - e.g. influenza virus comprises protein, RNA and lipid
- Viral nucleic acid encodes proteins for example for:
  - capsid
  - nucleic acid replication
  - virus-host interaction
- Viral nucleic acid does not code for:
  - ribosomes
  - mitochondria
- Therefore, viruses cannot carry out protein synthesis and energy production
  - dependent on host to provide these processes
  - explains the obligate nature of the parasitic interaction

Key steps in viral life-cycles

- Attachment to host cell
  - mediated by structures on virus particles
  - may target virus to particular cell type in a particular host
- Entry
  - attachment can induce changes in the structure of virus particle or viral proteins
  - these changes lead to entry of virus into cell or to fusion of viral envelope with cellular membrane
- Uncoating
  - release of nucleic acid into correct cellular compartment
  - may occur at the plasma membrane, in cytoplasm or at the nuclear membrane
- Expression of genetic information
  - production of proteins for:
    - virus particle
    - nucleic acid synthesis
    - virus-host interactions
  - may occur after replication of genetic material
- Replication of genetic material
  - DNA viruses often use cellular polymerase
  - RNA viruses generally encode their own polymerase
  - may occur before gene expression
- Assembly
  - viruses are assembled from pre-formed components
  - viruses do not reproduce by division
  - lipid envelope, if present, may be added at endoplasmic reticulum, Golgi apparatus or cell membrane
- Maturation
  - prepares virus for infection of the next cell
  - often involves proteolysis
  - may occur after release of the virus from the cell
- Release
  - may be accompanied by cell lysis (cytolytic viruses e.g. poliovirus)
  - may occur without cell lysis (non-cytolytic viruses e.g. hepatitis B virus – budding at plasma membrane)

1.2 Outline Baltimore's classification scheme for viruses. Give an example of a virus from each of the six classes.

Baltimore's classification scheme

- Baltimore's scheme uses relationship of viral genome to mRNA
  - information carried from genome to cellular ribosomes via mRNA
  - genetic information in viral genomes expressed as protein
- Applicable to all viruses because every virus must use host cell machinery for protein synthesis
  - every virus must generate mRNA from its genome
- Classification scheme eliminates much of variety found in viruses

Outline of Baltimore's classification scheme

- 6 classes of viral genome
  - positive sense strands of DNA (dotted lines) or RNA (full lines) in blue, negative in black
    - positive sense strands (also called plus or coding) have the same sense as the mRNA
    - negative sense strands (also called minus or non-coding) are complementary to the mRNA

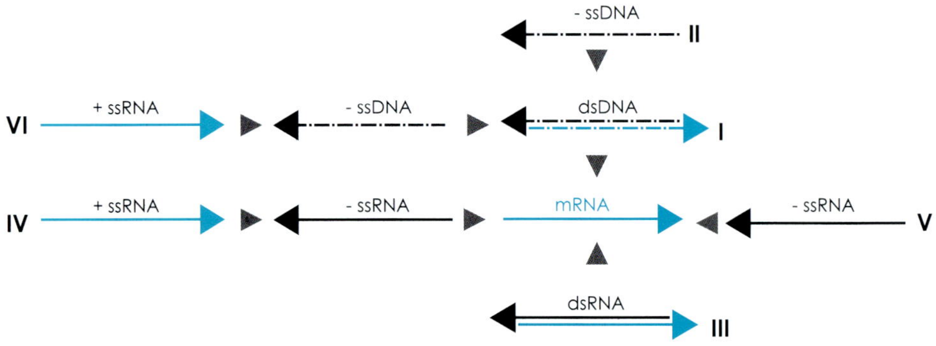

- Note the difference between class IV and class VI viruses
  - class VI genomic RNA is never used as mRNA, even though it is identical
  - class IV genomic RNA can be used as mRNA or copied into the minus sense from which much more positive sense RNA is generated

Virus examples

- Class I – double stranded DNA viruses
  - herpes simplex virus type 1
- Class II – single stranded DNA viruses
  - feline parvovirus
- Class III – double stranded RNA viruses
  - human rotavirus
- Class IV – single stranded positive sense RNA viruses
  - hepatitis C virus
- Class V – single stranded negative sense RNA viruses
  - measles virus
- Class VI – single stranded positive sense RNA viruses with a DNA intermediate (retroviruses)
  - human immunodeficiency virus type 1 (HIV-1)

1.3 Viruses keep their genomes small. Why? Give four strategies employed by viruses to achieve this goal.

Why do viruses keep their genome small?

- The larger the genome, the longer the time needed to replicate it
- The larger the genome, the larger the capsid must be
  - more coding capacity must be used to build the capsid
- The larger the genome, the higher the probability of mutations
  - for DNA viruses, use of cellular DNA polymerase can keep error rates low
    - cellular DNA polymerases have proofreading
  - for RNA viruses, RNA polymerases do not have proofreading
    - error rate approximately 1 in 10,000 nucleotides synthesised
    - genomes longer than 10,000 nucleotides would be synthesised with too many mutations to be viable
    - most RNA viral genomes are smaller than 10,000 nucleotides
    - exception: the RNA genome of coronaviruses (e.g. SARS virus) has about 30,000 nucleotides

Four strategies to keep viral genomes small

- All viruses use cellular machinery to generate energy and synthesise proteins
  - viral genomes do not need to code for such machinery
- Some viruses use the same region of nucleic acid to code for different proteins
  - reduces the amount of nucleic acid required
  - more than one open reading frame in same region of nucleic acid
  - either by using different reading frames on one strand e.g.
    - hepadnaviruses (hepatitis B virus),
    - retroviruses (human immunodeficiency virus 1)
  - or by encoding proteins on both strands
    - e.g. adenoviruses
- Some viruses produce one large RNA ("primary transcript") which is spliced by cellular splicing machinery to generate several mRNAs
  - only one set of signals for RNA synthesis
  - strategy of many DNA viruses, e.g.
    - adenoviruses
    - polyomaviruses (simian virus 40)
    - hepadnaviruses (hepatitis B virus)
  - but also found in RNA viruses, e.g.
    - retroviruses (HIV-1, human T-cell lymphotropic virus 1)
    - orthomyxoviruses (influenza virus)
- Some viruses produce single large protein (polyprotein) which is cleaved by viral and cellular proteases to mature proteins
  - only one mRNA required for several proteins
  - strategy of many positive stranded RNA viruses, e.g.
    - picornaviruses (poliovirus, hepatitis A virus)
    - flaviviruses (dengue virus, hepatitis C virus)
    - retroviruses (HIV-1, HTLV-1)

1.4 Describe the general evolution of the viruses. How do emerging viruses arise? Which parameters limit the evolution of viruses?

Evolution of viruses

- Two characteristic properties drive viral evolution
  - viruses produce a large number of descendants
  - viral replication generates large number of mutants
- Viruses can also evolve by recombination of genomes
- Certain segmented viruses (e.g. influenza virus, reovirus) can also evolve by reassortment of genome segments
- Large numbers of mutant viruses can be selected for fitness in new environments and hosts
  - new possibilities often provided by human behaviour, e.g.:
    - high density agriculture
    - water regulation
    - drug abuse
    - sexual practices

Emerging viruses are generated from existing ones

- Known virus changes to become able to infect an immune population
  - e.g. influenza virus
- Known virus is exposed to a naïve population
  - e.g. smallpox and measles viruses during the colonisation of the Americas
- Known virus changes its host-range to enter a different species
  - different species is by definition a naïve population
  - e.g. simian immunodeficiency virus (SIV) moving into humans as HIV-1

Which parameters limit the evolution of viruses?

- Icosahedral viral capsids limit size of nucleic acid
- Virus is dependent on host systems for decoding genetic information
- Mutations may adversely affect activity of viral proteins
- An extremely virulent virus kills its host and therefore itself
- Too many mutations lead to a non-viable virus
  - this concept is termed "error threshold"
  - RNA viruses replicate close to error threshold
  - DNA viruses replicate far from error threshold

Medical importance

- Compound ribavirin
  - can raise viral mutation rates over error threshold
  - used in combination with interferon α to treat hepatitis C virus infections

1.5 Name two types of symmetry found in virus particles. Give examples of virus families using these symmetries. What are the consequences of these two types of symmetry for the properties of the viruses?

Two types of symmetry found in virus particles:

- Icosahedral symmetry
    - regular polyhedron with 20 faces
        - closest regular shape to a sphere (largest volume, smallest surface)
    - polyhedron has three types of symmetry axes:
        - 12 five-fold, 20 three-fold and 30 two-fold axes
    - virus particle must have at least 60 copies of one protein to obey icosahedral symmetry
        - 60 is the smallest number divisible by 12, 20 and 30
    - icosahedral symmetry is found in:
        - non-enveloped and enveloped DNA and RNA viruses
        - but very infrequent in single stranded RNA viruses of negative sense
    - virus particle can have icosahedral symmetry without looking like an icosahedron
- Helical symmetry
    - proteins arranged in simple repeating unit around nucleic acid
    - each protein
        - occupies the same position
        - makes identical interactions with its neighbours and the nucleic acid
    - helical symmetry is found in:
        - many plant viruses
        - single stranded RNA animal viruses of negative sense
        - note important exception: coronaviruses are helical, enveloped viruses with RNA genome of positive sense

What are the consequences of these two types of symmetry?

- Icosahedral symmetry
  - internal volume of icosahedron limits size of nucleic acid which can be enclosed
  - including more, often different, proteins in the icosahedron increases the internal volume
    - rules of symmetry must still be obeyed
    - limits number of proteins to certain multiples of 60 – e.g. 180, 240, 420, 960
    - only possible if proteins make similar interactions with each other
    - proteins referred to as "quasi-equivalent"
  - differences in proteins on surface of icosahedron allows antigenic variation
- Helical symmetry
  - helical structure formed around nucleic acid
    - theoretically no limit to length
  - proteins occupying identical position makes antigenic variation difficult
    - most helical animal viruses have a lipid envelope
    - antigenic variation found in surface glycoproteins

1.6 Sketch the virus particle of a(n):
- non-enveloped RNA virus
- enveloped RNA virus
- non-enveloped DNA virus
- enveloped DNA virus.

Name at least one cellular molecule used by your selected viruses to attach to the host cell.

Non-enveloped RNA virus

- Poliovirus
  - member of picornavirus family
  - icosahedral virus with a single stranded RNA genome of positive sense
- Cellular molecule used by poliovirus for attachment:
  - protein of unknown function
  - member of immunoglobulin family

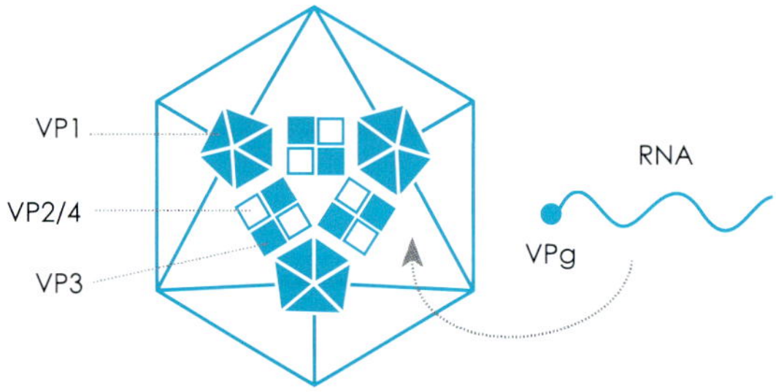

- RNA contained in particle has small virally encoded protein at 5′ end (VPg, viral protein genome-linked)

Enveloped RNA virus

- Rabies virus
  - member of rhabdovirus family
  - helical virus with a single stranded RNA genome of negative sense
- Cellular molecules used by rabies virus for attachment
  - include at least 3 proteins found on nerve cells
  - one is acetyl-choline receptor

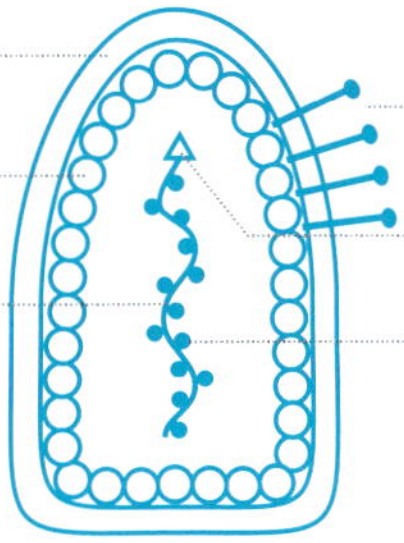

Non-enveloped DNA virus

- Simian virus 40 (SV40)
  - member of polyomavirus family
  - icosahedral virus with a circular double stranded DNA genome
- Cellular molecule used by SV 40 for attachment
  - major histocompatibility complex class I molecule

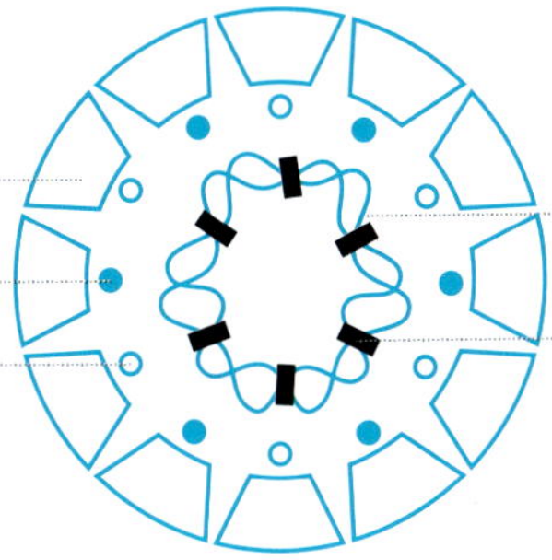

- SV40 DNA with bound nucleosomes is termed a minichromosome

Enveloped DNA virus

- Herpes simplex virus type I (HSV-1)
  - member of herpesvirus family
  - icosahedral virus with a linear double stranded DNA genome
- Cellular molecule used by HSV-1 for attachment
  - heparan sulphate

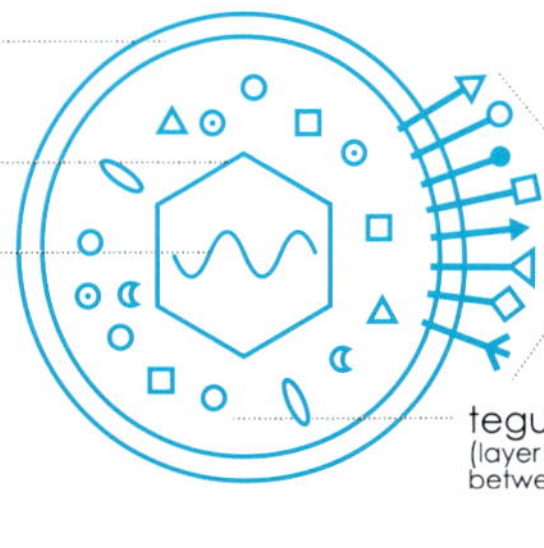

1.7 Give two examples of a virus which can be found in:
- blood
- faeces
- semen
- respiratory droplets and particles
- saliva
For each virus, indicate:
- which viruses (if any) are risk factors for cancer
- for which viruses (if any) a vaccine is available
- which viruses (if any) can cause long-term infections
- which viruses (if any) have a different course of infection depending on the infected person's age

BODY FLUID	VIRUS
blood	hepatitis B virus hepatitis C virus
faeces	human rotavirus poliovirus
semen	human immunodeficiency virus type 1 human T-cell lymphotropic virus type 1
respiratory droplets	measles virus influenza A virus
saliva	herpes simplex virus type 1 Epstein-Barr virus

- Presence of a virus in the blood is termed "viraemia"

Viruses which are risk factors for cancer

- Hepatitis B virus
- Hepatitis C virus
- Human T-cell lymphotropic virus type 1
- Epstein-Barr virus

- Hepatitis B virus
- Measles virus
- Influenza A virus
- Poliovirus
- Human rotavirus (major cause of infant diarrhoea worldwide)

Viruses which can cause long-term infections

- Hepatitis B virus
- Hepatitis C virus
- Herpes simplex virus type 1
- Human immunodeficiency virus type 1
- Human T-cell lymphotropic virus type 1
- Epstein-Barr virus

Viruses which have a different course of infection depending on the infected person's age

- Hepatitis B virus
  - infection in first years of life frequently leads to chronic infection
  - infection as adult usually cleared by immune system – low chance of chronic infection
- Epstein-Barr virus
  - infection in first years of life usually subclinical
  - infection as adolescent or young adult leads to infectious mononucleosis ("kissing disease")
  - infection with Epstein-Barr virus as child is however one of the risk factors for Burkitt's lymphoma

1.8 What is a virus? Which factors made it possible to eradicate smallpox virus?

What is a virus?

- A virus is an obligate intracellular parasite
  - obligate = has no choice
- All viruses contain nucleic acid and protein
  - enveloped viruses also contain lipid bilayer
- Viruses do not reproduce by division
  - one cannot say that viruses grow
- Instead, viruses are assembled from pre-formed components
  - viruses are said to replicate
- Viruses have no ribosomes and no mitochondria
  - therefore, viruses are different from intracellular bacterial parasites (e.g. Chlamydia)

Medical importance

- Replication of a virus in a host organism may lead to disease
  - virus may kill cell by lysis
    - virus termed cytolytic
  - viral infection may induce cell to kill itself
    - process termed apoptosis
- Pathogenicity (= disease-causing) often an accident
  - e.g. destruction of hepatocytes in hepatitis B and C virus infections
    - not a direct result of viral replication
    - host immune system eliminates infected cells
  - e.g. replication of a virus in an "unnatural host"
    - tick-borne encephalitis virus
    - no disease in rodents, the natural host
    - this virus may however cause encephalitis when it infects a human
- Ability of virus to infect a different host is one example of viral evolution

Eradication of smallpox virus

- Smallpox virus
  - member of poxvirus family
  - enveloped double stranded DNA virus with a complex virus particle
- Eradicated from the planet in 1977

Factors leading to eradication of smallpox virus

- Human only host
  - no animal reservoir
- One known serotype, antigenically stable
- Infected person cannot transmit virus until symptoms are visible
- No persistent infection

Vaccine

- Visible scar tissue from vaccination eliminates need for paper records
- Live vaccine gives life-long protection after single injection
- However, origin of vaccine strain (vaccinia virus) unclear
  - not directly related to cowpox virus used by Jenner
  - not a variant of smallpox virus
- Nevertheless, vaccine strains very similar all over the world

Economics of eradication

- Strong global will to eradicate virus
- Eradication cost much less than annual costs of:
  - quarantine
  - bureaucracy
  - travel restrictions

1.9 Which properties should a viral vaccine have to make it safe and efficient? How can killed viral vaccines and subunit vaccines be effective?

Which properties should a viral vaccine have to make it safe and efficient?

- A vaccine should induce protection against a pathogenic virus without itself causing disease
- Immune response generated by vaccine should resemble as closely as possible the response generated by pathogenic agent
  - e.g. immune response generated by measles virus vaccine is not as powerful as that generated by wild-type viral infection
    - reason not clear, as measles virus vaccine is a live viral vaccine
    - possibly as a result of the attenuation process
- Vaccines should induce the production of memory B and memory T cells
  - these cells can be rapidly reactivated when the pathogenic agent is encountered
  - memory B cells seem to contribute most to protective immunity
    - enable rapid production of correct neutralising antibodies
- Vaccines should induce long-term protection with just one or few inoculations
  - oral administration (e.g. oral poliovirus vaccine) should always be a goal
- Vaccines should be cheap
- Society should see a clear advantage in vaccine use

How can killed viral vaccines and subunit vaccines be effective?

- Killed and subunit vaccines cannot produce replicating virus
  - therefore, they cannot generate memory T cells
- However, killed and subunit viral vaccines do elicit a B cell response
  - virus particles or subunits taken up and degraded by dendritic cells and macrophages
    - peptides presented to helper T cells which activate B cells specific for the vaccine
  - most B cells develop into short-lived antibody producing plasma cells
  - some however develop into long-lived memory B cells
- Example: hepatitis B virus (HBV) vaccine
  - HBV vaccine employs just surface glycoprotein of HBV
    - prevention of disease by generation of memory B cells
  - in contrast, during a genuine HBV infection:
    - cytotoxic T cells, which destroy infected liver cells, are the primary defence against HBV
    - large scale synthesis of HBV specific antibodies does take place
    - but HBV produces vast quantities of virus-related particles which titre out antibodies
    - antibodies cannot neutralise HBV before it infects another liver cell
    - antibodies to HBV are measurable only after infection has been cleared

1.10 What are prions? Which diseases do they cause? How can they be transmitted? Which other factor can influence the outcome of these diseases?

What are prions?

- Prions are cellular proteins which can exist in two forms
  - normal, non-pathogenic, protease-sensitive form – termed PrP^{c} (c = cellular)
  - abnormal, pathogenic, protease-resistant form – termed PrP^{sc} (protease resistant protein found in scrapie)
- PrP^{c} and PrP^{sc} differ in their secondary and tertiary structure
  - PrP^{c} undergoes very slow, irreversible conformational change to PrP^{sc}
  - PrP^{sc} catalyses conversion of PrP^{c} to PrP^{sc}
    - the more PrP^{sc} present, the faster the conversion
- Note that it has not been proved conclusively that prions comprise exclusively proteins
  - purified preparations of prions may contain traces of nucleic acid or other macromolecules

Which diseases do prions cause?

- Prions are the causative agents of transmissible spongiform encephalopathies (TSEs)
- TSEs include:
  - scrapie (sheep)
  - bovine spongiform encephalopathy (BSE, cattle)
  - Creutzfeld-Jakob disease (CJD, humans)
  - new-variant Creutzfeld-Jakob disease (vCJD, humans)

How can prions be transmitted?

- TSEs thought to originate as genetic disease
  - genetic background of an animal or somatic mutation during life can cause conversion of PrP^{c} to PrP^{sc}
- Infectious prions can also be transmitted to another individual by e.g.:
  - sheep eating grass contaminated by placenta from sheep with scrapie
  - cattle eating feed contaminated with material from cattle with BSE
  - humans eating meat from cattle with BSE
  - human receiving growth hormone preparations from humans with CJD
- Transmitted PrP^{sc} catalyses conversion of PrP^{c} to PrP^{sc} in new host

Which other factor can influence the outcome of TSEs?

- Genetic background of recipient of infected material
  - e.g. in humans, amino acid 129 in gene encoding PrP^{c} influences development of vCJD
  - almost all cases of vCJD in persons homozygous for methionine at residue 129
    - reason: perhaps a more rapid conversion of PrP^{c} to PrP^{sc}
  - individuals heterozygous at residue 129 have lower rates of disease progression
  - genetic data suggest that heterozygotes have a selective advantage
    - perhaps because human populations were cannibalistic
    - humans homozygous for methionine at residue 129 may have developed disease before reproductive age

1.11 Name two cellular molecules used by human pathogens to enter a cell. Give three examples of sites at which viruses can uncoat. Use specific examples of viruses to illustrate your answer.

Two cellular molecules used by human pathogens to enter a cell

- Intracellular adhesion molecule 1 (ICAM-1)
  - used by majority of human rhinoviruses (common cold viruses)
  - member of the immunoglobulin family of proteins
  - functions in the host in lymphocyte communication
- Integrins
  - used by foot-and-mouth disease virus
  - function in the host by binding fibronectin (extracellular matrix protein)
  - important interaction in maintaining extracellular matrix

Uncoating of the viruses

- At the cell membrane - e.g. human immunodeficiency virus (HIV-1)
  - HIV-1 binds to its cellular receptors CD4 and CCR5 or CXCR4 using its envelope proteins gp120 and gp41 (products of *env* gene)
    - gp120 binds first to CD4, causing gp120 to change conformation
    - conformational change exposes gp120 binding site for CCR5
    - gp120 binding to CCR5 causes conformational change in gp41
  - change in gp41 exposes membrane fusion peptide
    - fusion peptide inserts into cell membrane
    - viral envelope and cell membrane fuse
  - contents of virus particle released into cytoplasm
- At the endosome - e.g. influenza virus
  - influenza virus envelope protein haemagglutinin (HA) binds sialic acid on cellular surfaces
    - virus taken up by endocytosis into endosome
    - low pH in endosome causes conformational change in HA
  - membrane fusion peptide exposed
    - fusion peptide inserts into endosomal membrane
    - viral envelope and endosomal membrane fuse
  - viral RNA genome released into cytoplasm
    - nucleoprotein bound to RNA genome mediates import of the genome into nucleus
- At the nuclear membrane - e.g. adenovirus
  - adenoviruses bind to two cellular proteins
  - binding to one molecule allows virus to attach to cell
    - e.g. certain adenoviruses use major histocompatibility complex class I molecule
  - binding to a second allows virus taken up by endocytosis into endosome
    - e.g. certain adenoviruses use a member of integrin protein family
    - low pH in endosome causes conformational changes and loss of certain proteins from adenovirus particle
  - endosome is lysed, possibly by action of released viral proteins
    - remaining proteins in virus particle bind to nuclear membrane
  - virus uncoats releasing viral DNA into nucleus

1.12 What are the functions of the virus particle? Why are virus particles described as being metastable? Give an example how a metastable virus particle can be generated using proteolytic cleavage.

Functions of the virus particle

- Virus particles comprise
  - genetic material (DNA or RNA)
  - proteins
  - lipid bilayer in enveloped viruses
- Virus particles
  - protect viral genetic material from the environment
  - introduce and release viral genetic material into a suitable cell
  - may provide enzymes for:
    - essential steps in virus life-cycle
    - modulation of cellular physiology

Why are virus particles metastable?

- Metastable particles
  - are particles which have not yet reached their lowest energy level
  - the transition to the lower energy level can be induced by specific biological interactions
    - such as binding of virus to receptor - e.g. HIV-1
    - lowering of pH in endosomes - e.g. influenza virus
  - leads to correct release of genetic material
- Metastability allows virus particles to perform the seemingly paradoxical functions of protection and targeted release of genetic material

How can a metastable particle be generated?

- Metastable particles are generally produced during synthesis and assembly
- One common mechanism is proteolysis of a viral protein
  - e.g. in influenza virus, cleavage of haemagglutinin (HA0) into HA1 and HA2
  - e.g. in HIV-1, cleavage of gp160 (*env* gene product) into gp120 (surface, SU) and gp41 (transmembrane, TM) proteins

Effects of proteolytic cleavage

- E.g. influenza virus (see diagram)
- Proteolytic cleavage turns one polypeptide chain into two (drawings a and b below)
  - two chains remain together, attached through a disulphide bridge
- Newly generated N-terminus important for membrane fusion
  - "fusion peptide" present at new N-terminus
- Fusion peptide remains hidden until exposed by conformational changes
  - induced by lowering of pH (drawing c below)
  - fusion peptide inserts itself into host cell membrane
    - brings viral and cellular membranes together for fusion

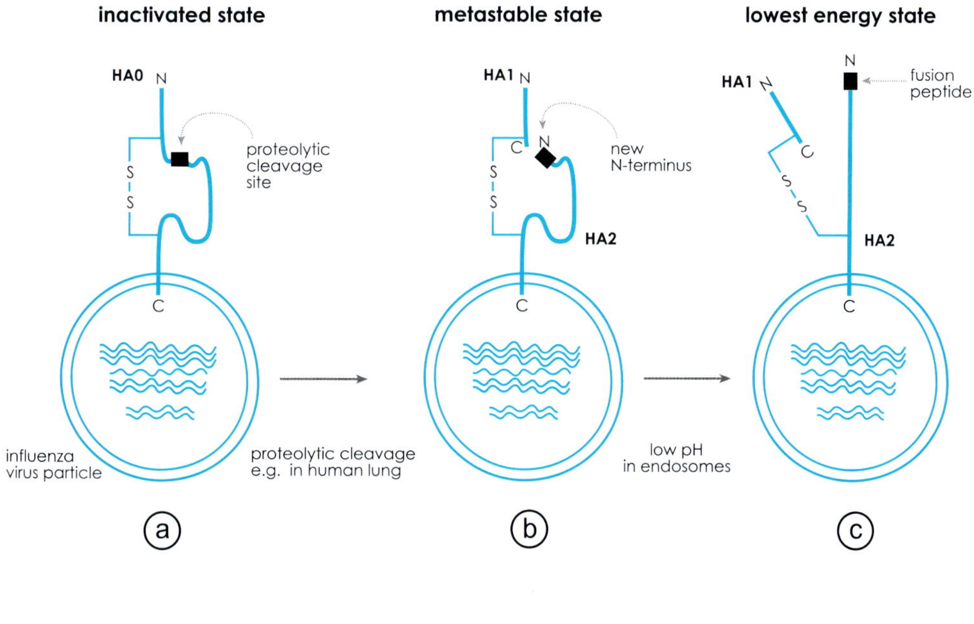

1.13 Why do many DNA viruses encode proteins which influence the cell cycle and apoptosis? What consequences may the activity of these proteins have for the infected cell?

Why do many DNA viruses encode proteins which influence the cell cycle and apoptosis?

- DNA viruses often depend on host cell machinery for synthesis of their DNA genome
    - however, most animal cells are in G_1- or G_0-phase of the cell cycle
        - G_1 and G_0 are phases of the cell cycle which do not synthesise DNA
- For this reason, many DNA viruses encode a protein which causes the cell to leave G_1- or G_0-phase and enter S-phase
    - S-phase is DNA synthesis phase
- Host cell has mechanisms that sense that DNA synthesis is not scheduled
    - e.g. cell size may be recognised as too small
- unscheduled DNA synthesis is generally a signal for a cell to enter apoptosis
    - i.e. cell kills itself by programmed cell death
- For this reason, many DNA viruses encode a protein which prevents apoptosis
- Examples of DNA viruses encoding proteins which influence the cell cycle and apoptosis:

	proteins influencing	
	cell cycle	apoptosis
simian virus 40 (SV40)	large T antigen	large T antigen
human papillomavirus	protein E7	protein E6
adenovirus	protein E1A	protein E1B

What consequences may the activity of these proteins have for the infected cell?

- Expression of these proteins may be one step towards cellular transformation
- E.g. human papillomavirus (HPV)
  - most infections with this virus are cleared by immune system
  - however, very occasionally a very small fraction persists
  - HPV DNA is no longer episomal
    - instead some fragments are integrated into host chromosome
    - leads to higher expression of E6 and E7 proteins
    - together with other mutations affects growth properties of the cells

- Concepts
  - transformation requires expression of viral DNA, but not virus production
  - transformation is a result of viral protein activity, but does not offer advantages to the virus

2 RNA viruses

2.1 Name two families of viruses with a positive sense RNA genome. Sketch their genome structures and point out important similarities and differences. Compare and contrast their strategies for expression of genetic information.

Picornavirus

- Non-enveloped, icosahedral viruses with a single stranded RNA genome of positive sense
- Members include:
  - poliovirus
  - foot-and-mouth disease virus
  - hepatitis A virus

Togavirus

- Enveloped, icosahedral viruses with a single stranded RNA genome of positive sense
- Members include:
  - Semliki Forest virus
  - Sindbis virus
  - rubella virus

Genome structure

- Picornavirus

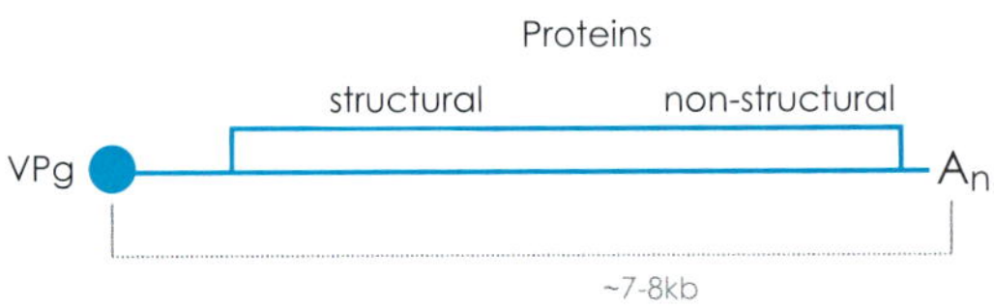

- Togavirus

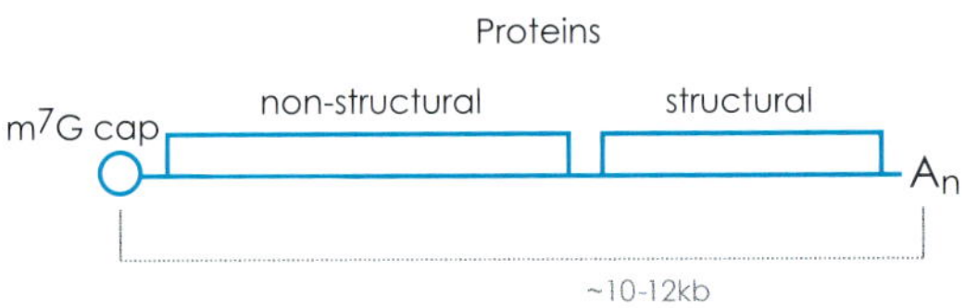

- Similarities:
  - both RNA genomes have positive sense and polyA sequence
- Differences:
  - picornaviruses have viral protein (VPg) linked to the 5' end of RNA whereas togaviruses have 5' cap structure
  - picornaviruses have just one open reading frame (ORF) whilst togaviruses have two
  - 5' region of picornavirus genome encodes structural proteins
  - 5' region of togavirus genome encodes non-structural proteins

Strategies for expression of genetic information

- Similarities:
  - replication of both viruses takes place in cytoplasm
  - both viral RNAs used for protein synthesis on release into cytoplasm
  - both viruses use proteolytic processing
- Differences:
  - picornaviruses produce one single polyprotein
    - cleaved by viral proteases to mature proteins
    - no glycosylation of proteins
  - togaviruses produce proteins from two ORFs
    - proteins from 5' ORF cleaved by viral protease
    - proteins from 3' ORF cleaved by part of capsid protein which can act as a protease and by cellular proteases
    - glycosylation of the envelope protein takes place in the endoplasmatic reticulum

2.2 Name two families of viruses with a positive sense RNA genome not mentioned in question 2.1. Sketch their genome structures and point out important similarities and differences. Compare and contrast their strategies for expression of genetic information.

Flavivirus

- enveloped, icosahedral viruses with a single stranded RNA genome of positive sense
- members include:
  - hepatitis C virus
  - yellow fever virus
  - tick-borne encephalitis virus

Coronavirus

- enveloped, helical viruses with a single stranded RNA genome of positive sense
- members include:
  - SARS virus
  - mouse hepatitis virus

Genome structure

- Flavivirus

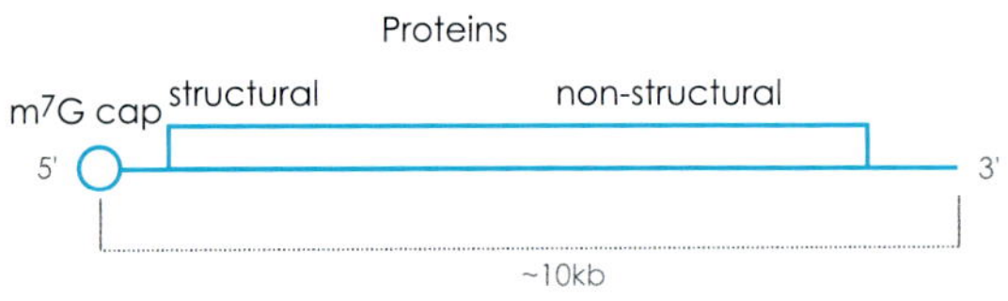

- Coronavirus

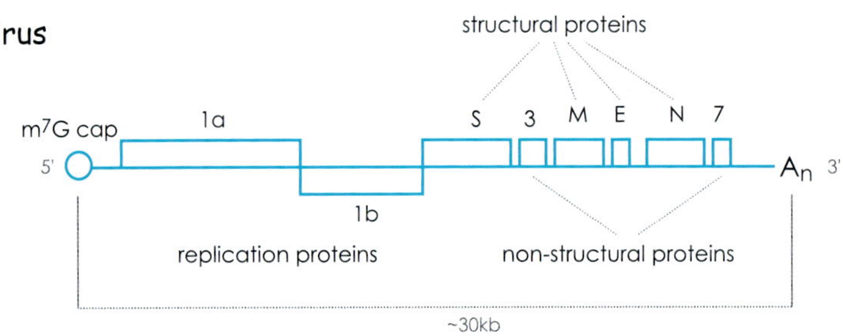

- Similarities
  - both RNAs have positive sense and a 5' cap structure
- Differences
  - flaviviruses have just one open reading frame (ORF), whereas coronaviruses have several
  - 5' region of flavivirus genome encodes structural proteins
  - 5' region of coronaviral genome encodes proteins for replication
  - only coronaviral RNA has a polyA sequence

Strategies of expression of genetic information

- Similarities
  - replication of both viruses takes place in cytoplasm
  - both viral RNAs used for protein synthesis on release into cytoplasm
  - both viruses use proteolytic processing
  - both viruses glycosylate certain proteins in endoplasmic reticulum
- Differences
  - flaviviruses produce only one single protein
    - cleaved by cellular and viral proteases to mature proteins
  - coronaviruses produce proteins from several open reading frames
    - initial translation from one AUG start codon produces two polyproteins
    - ribosomal frame-shifting allows protein synthesis from second open reading frame
    - proteolytic processing of these two proteins gives mature proteins including polymerase
    - for remaining open reading frames, RNA polymerase makes sub-genomic negative sense RNAs corresponding to required open reading frame
    - sub-genomic negative sense RNAs copied to sub-genomic mRNAs
    - mRNAs translated to give second set of proteins

2.3 To which virus family does hepatitis C virus belong? How is hepatitis C virus transmitted? How can infections with this virus be treated? Why is it difficult to develop a vaccine against this virus?

Hepatitis C virus (HCV)

- Member of flavivirus family
- Enveloped, icosahedral virus with a single stranded RNA genome of positive sense
- Often causes chronic infection of hepatocytes
- Only host of HCV is man
- In contrast to other viruses of this family and their often multiple hosts:

VIRUS	HOSTS
yellow fever virus	man, monkey, mosquitoes
dengue virus	man, monkey, mosquitoes
West Nile virus	man, birds, horses, mosquitoes

Transmission of HCV

- HCV transmitted via:
  - contaminated blood and blood products
  - contaminated needles in drug abuse and body decorations
  - contact between maternal and infant blood during child birth
- HCV not transmitted to any significant extent by sex
- HCV not transmitted by arthropod vectors

Treatment of HCV infections

- HCV infections are treated with ribavirin and interferon α (IFNα)
  - ribavirin is a nucleotide analogue that can be incorporated into viral RNA
    - pairs with either cytidine or uridine, leading to a higher rate of mutation
  - virus mutation rate exceeds error threshold
    - i.e. too many mutations introduced leading to non-viable virus
  - IFNα mimics innate immune response
    - e.g. by stopping protein synthesis in the infected cell
- Treatment only effective in 20% to 40% of cases
  - some genotypes do not respond
  - depends on properties of viral protein NS5A
  - people over 40 respond poorly

Why has a vaccine against HCV not been developed?

- An in vitro culture system has only recently been developed
  - but such a system not required for development of hepatitis B virus vaccine!
- Presence of several different HCV genotypes
  - vaccine required which is effective against all genotypes
- Rapid evolution of virus during infection
  - HCV RNA polymerase frequently makes errors
    - enzyme has no proofreading ability
  - permanent introduction of new mutations
  - vaccine must protect against transmission of newly evolved viruses
- How can a vaccine be developed against viruses which have yet to evolve?

2.4 Name two members of the rhabdovirus family. Sketch the rhabdovirus genome structure and outline how the genetic information is expressed.

Rhabdoviruses

- Enveloped, helical viruses with a single stranded RNA genome of negative sense
- Members of the rhabdovirus family include:
  - rabies virus
    - important human pathogen
    - worldwide many thousand cases per year
  - vesicular stomatitis virus (VSV)
    - important pathogen of farm animals such as cattle, horses and pigs
    - causes lesions in mouth and teats

Rhabdovirus genome

- RNA possesses five genes as well as a non-coding leader region
  - N - nucleocapsid protein
  - M - matrix protein
  - L and P together make up the RNA polymerase
  - G - surface glycoprotein

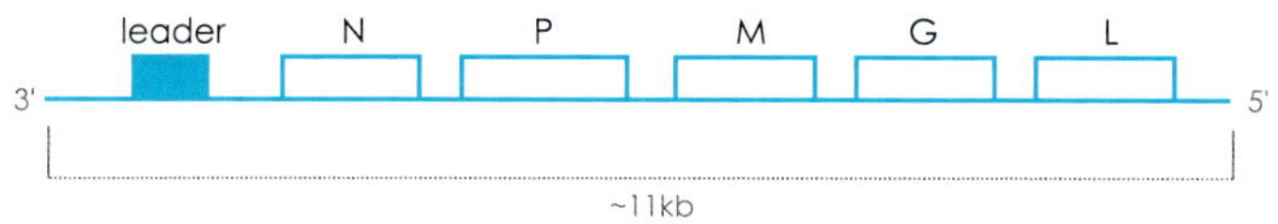

- RNA genome has negative sense
  - cannot therefore be translated directly into protein

Expression of genetic information in rhabdoviruses

- Virus particle contains viral RNA polymerase, a complex of proteins L and P
  - negative sense genomic RNA copied into positive sense sub-genomic RNAs on release into cytoplasm
- However, not all genes are expressed to the same extent
  - viral RNA polymerase synthesises more mRNAs corresponding to 3' end of the genomic RNA than 5' end
  - much more mRNA for N protein than for P protein
  - phenomenon known as transcriptional polarity

Mechanism of transcriptional polarity

- RNA polymerase initiates RNA synthesis at 3' end
- At the end of each gene sequence, RNA polymerase stops
  - RNA polymerase can either reinitiate synthesis or dissociate from RNA
    - therefore not all RNA polymerase molecules reach 5' end of RNA genome
  - result is fewer mRNAs from 5' end

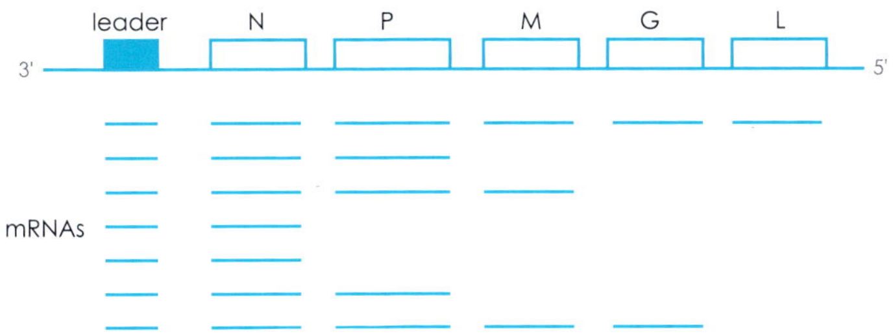

- Similar mechanisms found in paramyxoviruses (e.g. measles virus) and filoviruses (e.g. Ebola virus)

2.5 Name one RNA virus with a negative sense genome and one with a positive sense genome. Sketch their genome structures. Compare and contrast their strategies for the expression of genetic material.

Measles virus

- Member of paramyxovirus family
- Enveloped, helical virus with a single stranded RNA genome of negative sense

Hepatitis C virus

- Member of flavivirus family
- Enveloped, icosahedral virus with a single stranded RNA genome of positive sense

Genome structure

- Measles virus
  - has six open reading frames

- Hepatitis C virus
  - has one open reading frame

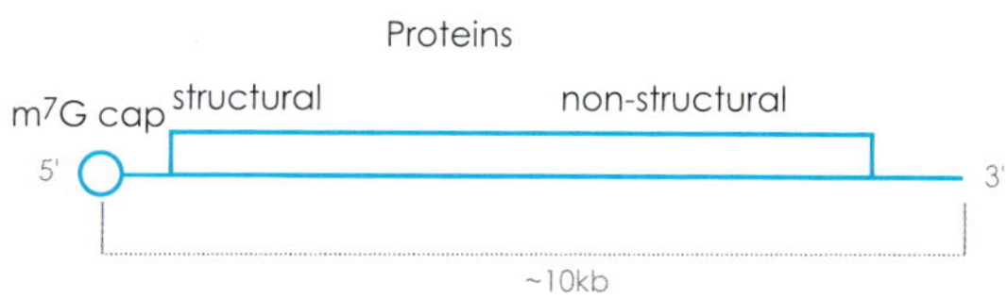

Strategies for expression of genetic material

- Measles virus
  - measles virus RNA cannot be directly translated
    - instead, RNA polymerase is present in virus particle
    - makes positive sense sub-genomic copies (i.e. mRNAs) of RNAs encoding measles virus mature proteins
    - mRNAs have 5' cap structure and 3' poly A tail
    - viral mRNAs are translated as normal cellular mRNAs
  - makes no polyprotein and does not encode proteases
  - proteins are produced in different amounts
    - RNA polymerase synthesises more RNAs corresponding to 3' end of the genome
    - sequence around initiating AUG codon can also influence amount of protein synthesised
  - glycosylation of envelope protein
    - like cellular glycoproteins, viral glycoproteins travel through endoplasmic reticulum and Golgi apparatus

- Hepatitis C virus (HCV)
  - in contrast to measles virus, HCV RNA is translated directly into protein
    - translation initiates on internal ribosome entry site (IRES)
  - translation generates single large polyprotein
    - cleaved by viral and cellular proteases into mature proteins
    - mature proteins produced in equal amounts
  - glycosylation of envelope protein
    - part of polyprotein containing glycoprotein enters and then leaves lumen of endoplasmic reticulum
    - following proteolytic cleavage, envelope protein matures through Golgi apparatus like cellular glycoproteins

2.6 Explain in outline the replicative cycle of measles virus. How is measles virus transmitted and how does it deal with the host immune system?

Measles virus

- Member of paramyxovirus family
- Enveloped, helical virus with a single stranded RNA genome of negative sense

The replicative cycle of measles virus

- Measles virus binds to receptor and uncoats at cell membrane
  - identity of receptor molecule used in vivo unclear
- Viral RNA polymerase present in particle
  - immediately makes sub-genomic positive mRNAs corresponding to regions encoding viral proteins
  - viral proteins synthesised using cellular systems
  - when enough nucleocapsid protein is present, viral RNA polymerase synthesises full-length positive strand RNA
  - full-length negative strand genomic RNA is than made from the full-length positive RNA template
- New measles virus particles bud from the membrane
  - contain genomic RNA, nucleocapsid and RNA polymerase
  - envelope of virus has viral proteins required for cell entry and fusion of virus and cell membranes

Transmission of measles virus

- Via respiratory droplets
  - exhaled from lungs via coughing
- Although measles virus enters and leaves body via lungs, infection is systemic

Measles virus and the immune system

- Measles virus has only one serotype
  - protein which binds to cellular receptor is targeted by antibodies
  - changes in this protein at the site which induce antibodies prevent receptor binding
    - therefore protein cannot change like haemagglutinin of influenza virus
    - no antigenic drift in measles virus
- Measles virus interferes with innate immunity by unknown method
- Measles virus interferes with adaptive immunity by productively infecting dendritic cells (DCs)
  - virus initially infects respiratory epithelial cells
  - innate immunity fights infection, causes DCs to migrate to lymph nodes
  - other white blood cells also infected, including macrophages
    - infection of DCs and macrophages suppresses immune system
- Virus produced in lymph nodes has time to return to lung
  - re-infection of respiratory epithelial cells occurs
  - immunosuppression allows high levels of virus to be produced and then exhaled from lung

2.7 Name an RNA virus with a negative sense genome and one with a positive sense genome not mentioned in question 2.5. How do the life-cycles of these viruses differ?

Poliovirus

- Member of picornavirus family
- Non-enveloped, icosahedral virus with a single stranded RNA genome of positive sense

Vesicular stomatitis virus

- Member of rhabdovirus family
- Enveloped, helical virus with a single stranded RNA genome of negative sense

Poliovirus life-cycle

- Genome of poliovirus

  - single open reading frame (ORF), two untranslated regions (UTR)
    - viral RNA polymerase not present in virus particle
    - RNA genome linked to viral protein, VPg
  - on entry, cellular ribosomes translate RNA into single polyprotein
  - polyprotein cleaved by virally encoded proteases to mature proteins including viral RNA polymerase
- Genomic RNA synthesis
  - VPg primes viral RNA polymerase to synthesise negative sense RNA
  - generates double stranded (ds) RNA intermediate (called replicative intermediate)
    - new positive sense copies made from ds RNA intermediate

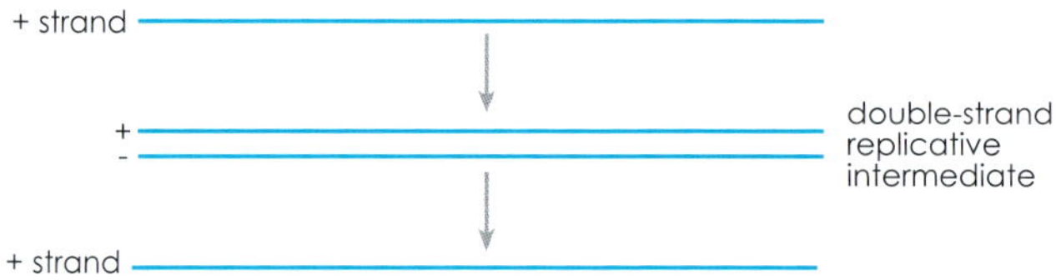

- New positive sense RNAs can be:
  - translated to make viral proteins
  - packaged with capsid proteins to make new virus particles

Vesicular stomatitis virus life-cycle

- Genome of vesicular stomatitis virus

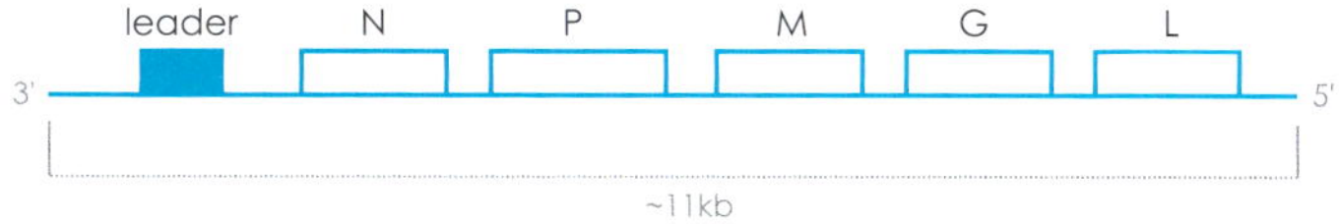

  - 5 open reading frames, one non-coding RNA (leader)
  - viral RNA polymerase (proteins L & P) present in virus particle, RNA bound by nucleocapsid (N)
  - on release into cytoplasm, negative sense RNA copied into sub-genomic positive sense mRNAs
  - sub-genomic positive mRNAs capped, polyadenylated and translated into viral proteins
    - newly synthesised N protein binds to input genomic RNA
- Genomic RNA synthesis
  - when sufficient N protein bound to input RNA, viral polymerase synthesises full length positive sense copies
  - generates double stranded (ds) RNA intermediate (called replicative intermediate)
    - new negative sense copies made from ds RNA intermediate

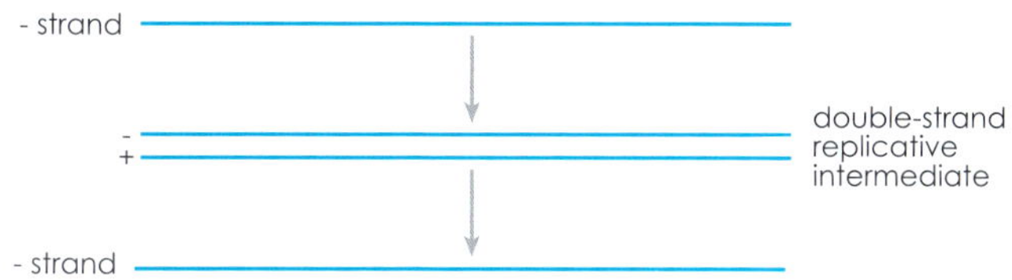

- New negative sense RNAs can be:
  - used for synthesis of sub-genomic mRNAs
  - packaged into virus particle with matrix (M) protein and lipid-bilayer containing viral glycoprotein (G)

Major differences in replication strategies

POLIOVIRUS	VESICULAR STOMATITIS VIRUS
• no polymerase in virus particle • RNA translated immediately • initial translation product processed by viral proteases • no requirement for nucleocapsid protein	• polymerase in virus particle • positive sub-genomic RNAs synthesised first • no proteolytic processing • full-length positive RNA produced when nucleocapsid protein concentration rises

2.8 Which structures are found on the surface of the influenza A virus? Explain how the RNA genome reaches the nucleus.

Influenza virus

- Member of the orthomyxovirus family
- Enveloped, helical virus with a segmented single stranded RNA genome of negative sense

Influenza A virus particle

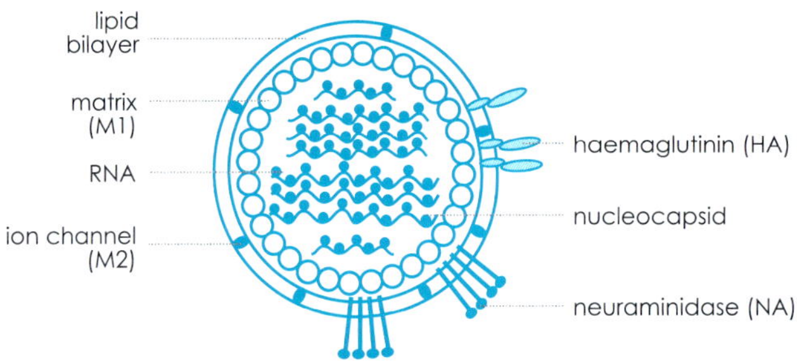

- Influenza A virus has eight genome segments

Surface structures

- Haemagglutinin (HA), 15 subtypes
- Neuraminidase (NA), 9 subtypes

Functions of HA and NA

- HA binds to sialic acid of glycoproteins on cell surface
  - HA induces fusion of endosomal and viral membrane
- NA cleaves sialic acid (N-acetyl neuraminic acid) from viral and cellular glycoproteins
  - NA allows influenza virus to be released from infected cells

Medical importance

- HA and NA determine antigenic properties of influenza A virus
- Inhibitors of NA used to treat influenza

How does the viral RNA reach the nucleus?

- Problem:
  - virus particle outside the cell
  - RNA surrounded by viral envelope
  - but viral RNA synthesis takes place in nucleus where the eight genomic seqments are copied into
    - sub-genomic mRNAs (positive sense)
    - new genomic RNAs (negative sense)
- Solution:
  - haemagglutinin (HA) on influenza virus binds to sialic acid of cell surface glycoproteins
  - complex of virus bound to sialic acid internalised into endosome
    - H^+ pumped into endosome, endosomal pH reaches pH 5
  - at pH 5, HA undergoes massive structural rearrangements
    - fusion peptide of HA inserts into endosomal membrane
  - H^+ pumped into interior of virus via ion channel protein M2
    - viral gene segments (as RNA-protein complexes) freed from matrix protein M1
  - viral and endosomal membranes fuse
    - viral gene segments are released as RNA-protein complex
  - nuclear localisation signal enable transport of RNA-protein complexes into nucleus
- The key
  - HA is a metastable protein
    - such a structure has yet to reach its lowest energy level
    - structural rearrangements during viral assembly generate metastable state
    - in human influenza virus, extracellular protease in the lung normally induces such rearrangements
  - energy for structural rearrangements derived from peptide bond cleavage

2.9 Compare and contrast the antigenic properties of hepatitis A virus and influenza A virus. How do the differences affect vaccination strategies? What are the differences between influenza A viruses and how do they arise?

The contrasting antigenic properties of the viruses are responsible for different vaccination strategies

- Hepatitis A virus (HAV)
  - member of the picornavirus family
  - non-enveloped, icosahedral virus with a single stranded RNA genome of positive sense
  - only one serotype which is antigenically stable
    - virus does not change with time
    - vaccine can protect over a long time period
  - HAV vaccine is an inactivated virus vaccine
    - protection is not life-long and must be reinforced every ten years
    - antibodies and memory cells decline with time
- Influenza A virus
  - member of the orthomyxovirus family
  - enveloped, helical virus with a segmented single stranded RNA genome of negative sense
  - in contrast to HAV, new i.e. antigenically different influenza A viruses appear each year
    - vaccine from last year may not protect against this year's viruses
    - each year influenza virus vaccine is reformulated to protect against currently circulating influenza A viruses

What are the differences between influenza A viruses?

- In influenza A virus, subtypes are determined by the antigenic properties of two viral envelope proteins:
  - haemagglutinin (HA)
  - neuraminidase (NA)
- 15 subtypes of HA and 9 subtypes of NA are known
  - influenza virus subtypes are designated e.g. H1N1, H1N2 depending on combination of HA and NA
  - only HA subtypes 1, 2 and 3 can be transmitted from human to human

Antigenic properties of influenza A virus can change in two ways

- Antigenic drift
  - error-prone RNA polymerase introduces mutations into HA and/or NA coding sequences
  - sufficient number of mutations during one year that vaccine strains no longer provide immunity against new ones
  - mutations do not however change:
    - the subtype of HA or NA
    - the ability of HA and NA to perform their functions
- Antigenic shift
  - influenza A virus strain acquires a different HA or NA sub-type
    - e.g. circulating subtypes H1N1 and H2N2 infect same cell
    - gene segments of influenza virus can be reassorted
    - as well as original virus, a new strain, e.g. H1N2, can be generated
    - new strain is termed reassortant
  - reassortant is new to human immune system
    - leads to worldwide outbreaks (pandemics) of influenza
    - this phenomenon gave rise to influenza A virus pandemics in 1957 and 1968

3 Retroviruses

3.1 Draw schematically the retrovirus particle. Which macromolecules are needed to transcribe the RNA genome into proviral DNA? How does the integration of the proviral DNA into the host genome occur?

Retrovirus family

- Enveloped, icosahedral viruses with a single stranded RNA genome of positive sense

Retrovirus particle

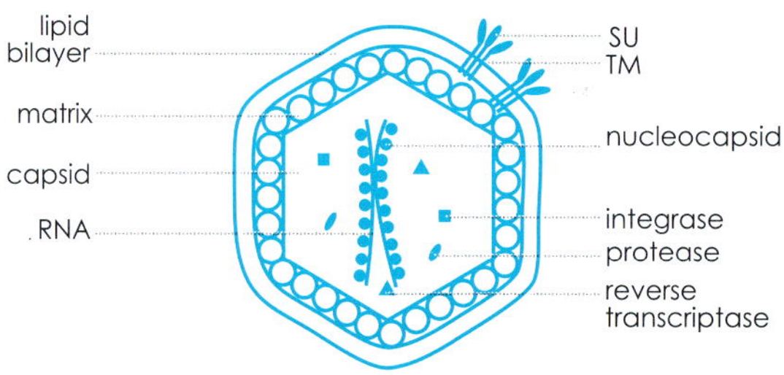

- Comments
    - cellular tRNA present, annealed to genomic RNA at specific site
    - two copies of genomic RNA
    - genomic RNA bound by nucleocapsid protein
    - *env* gene product cleaved during maturation into
        - transmembrane domain (TM) - contains fusion peptide
        - surface domain (SU) - binds to cellular receptor

Macromolecules required for proviral DNA synthesis

- Two copies of the retroviral genome
- Reverse transcriptase
- Cellular tRNA
  - tRNA binds to specific site on retroviral genome
    - termed primer binding site (PBS)
  - tRNA initiates the synthesis of DNA by reverse transcriptase
  - each retrovirus is primed by specific cellular tRNA
- Reverse transcriptase synthesises one double stranded DNA molecule (termed "proviral DNA") from two genomic RNA molecules

Integration of proviral DNA into host genome

- Proviral DNA and integrase form a complex
  - integrase removes two bases from proviral DNA
- Proviral DNA-integrase complex migrates to nucleus
  - ends of proviral DNA anneal to host DNA
    - host DNA cleaved and proviral DNA integrated
  - integration in the host chromosomal DNA is not site-specific
  - no energy input required
- Integration does not occur in resting (non-dividing) cells
  - proviral DNA-integrase complex too large to pass through pores in nuclear membrane
    - therefore, integration only occurs at mitosis when nuclear membrane is disassembled
  - exception: human immunodeficiency virus 1 (HIV-1) which encodes protein Vpr
    - allows proviral-integrase complex to cross nuclear membrane
- Integration gives rise to life-long infections
  - e.g. HIV-1, human T-cell lymphotropic virus 1 (HTLV-1)

3.2 Explain the overall layout of the retroviral genome. Why is the retroviral genome unique? State the properties of reverse transcriptase.

Retrovirus family

- Enveloped, icosahedral viruses with a single stranded RNA genome of positive sense
- RNA is replicated through a DNA intermediate
  - in contrast to all other RNA viruses
- Medical importance:
  - human immunodeficiency virus type 1 (HIV-1)
  - human T-cell lymphotropic virus type 1 (HTLV-1)

Genome structure of simple retrovirus (e.g. avian leukosis virus)

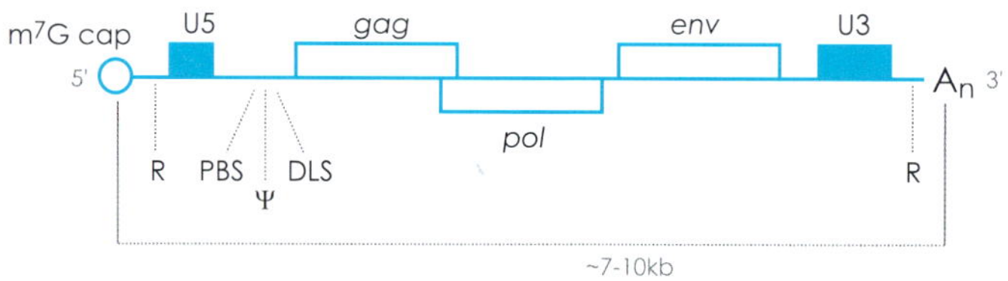

- Non-coding elements:
  - R - repeat sequences
  - U5 - unique 5' sequence
  - PBS - primer (tRNA) binding site
  - ψ - packaging sequence
  - DLS - dimer linking sequence
  - U3 - unique 3' sequence
- Coding sequences:
  - *gag* (group specific antigen)
    - nucleocapsid
    - matrix
    - capsid
    - protease
  - *pol*
    - reverse transcriptase
    - integrase
  - *env*
    - surface glycoprotein

Retroviral genome is unique because it is the

- Only viral RNA genome which is present in virus particle as two copies
    - can be termed "diploid"
- Only positive sense viral RNA not immediately translated into protein on entry into a cell
    - instead retroviral genome transcribed into proviral DNA by reverse transcriptase
- Only viral RNA which requires cellular tRNA to prime synthesis of intermediate
- Only viral RNA genome synthesised by cellular RNA polymerase

Properties of reverse transcriptase

- RNA dependent DNA polymerase
- DNA dependent DNA polymerase
- Degrades RNA component of RNA/DNA hybrid
- Possesses DNA unwinding activity
- Lacks 3'-5' proofreading activity
    - makes on average one error for every 10,000 bases synthesised
    - i.e. one error for every HIV-1 genome copied

3.3 How are the retroviral genomic RNA and the proviral DNA related? How is the defined retroviral RNA genome generated from proviral DNA?

Retroviruses

- Enveloped, icosahedral viruses with a single stranded RNA genome of positive sense

Unique features of retroviruses

- During viral replication, the viral enzyme reverse transcriptase copies genomic RNA into proviral DNA
  - proviral DNA integrated into host chromosomal DNA
  - host RNA synthesis and processing machinery generates viral genomic RNA

Retroviral RNA genome

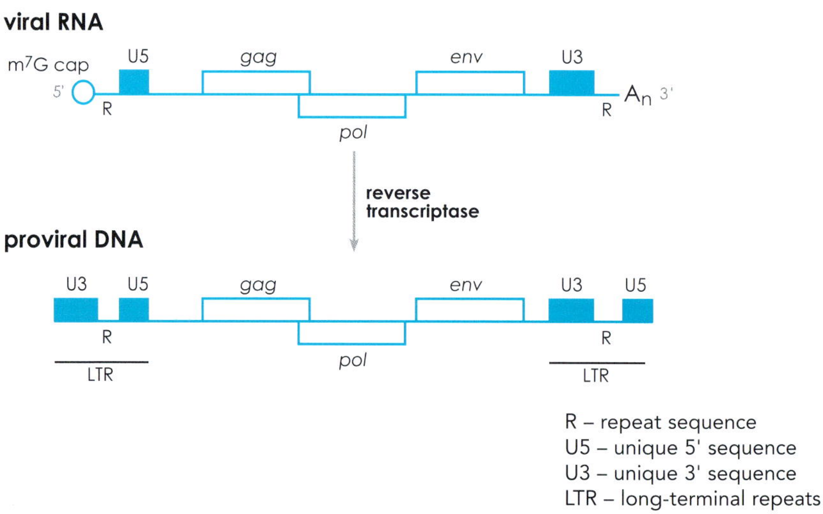

- Proviral DNA is longer than genomic RNA
  - "extra" copy of U3 at 5' end and U5 at 3' end
  - LTRs only present in proviral DNA

How is the defined retroviral RNA genome generated from proviral DNA?

- Host RNA polymerase II transcribes proviral DNA into RNA
  - signals in the proviral DNA recruit RNA polymerase II

Production of defined 5' end

- RNA synthesis starts exactly at first nucleotide of repeat sequence
  - alignment of promoter defines 5′ start of genomic RNA
- Promoter and other DNA elements in U3 may determine tissue or cell specific RNA expression
  - e.g. HIV-1 U3 has binding sites for proteins found in helper T cells

Production of the defined 3' end

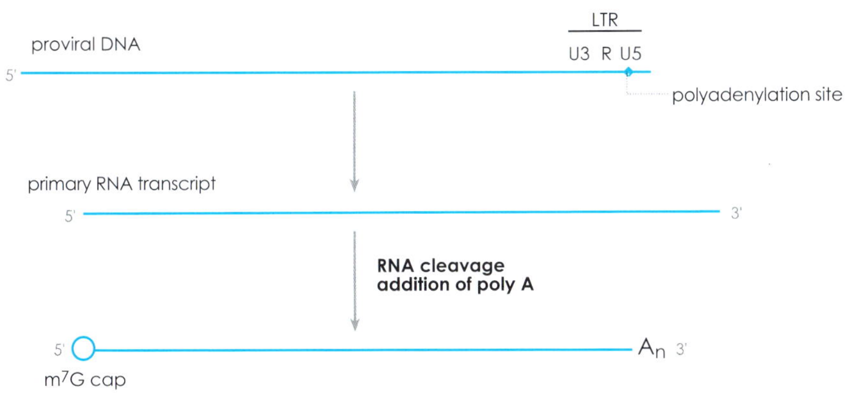

- Polyadenylation signal in U5 sequence allows RNA processing
  - RNA is cleaved and polyadenylated at last nucleotide of repeat sequence
  - position of polyadenylation signal defines 3′ end of genomic RNA

3.4 Explain the essential steps leading to production and integration of retroviral proviral DNA in the host chromosome. What are the consequences of this integration?

Retrovirus family

- Enveloped, icosahedral viruses with a single stranded RNA genome of positive sense
- Two copies of RNA genome in virus particle
- Members include:
  - avian leukosis virus (ALV)
  - human immunodeficiency virus type 1 (HIV-1)
  - human T-cell lymphotropic virus type 1 (HTLV-1)

Production of proviral DNA

- Retrovirus binds to specific proteins on cell surface
  - e.g. ALV binds to low density lipoprotein receptor
- Viral envelope and cell membrane fuse
  - contents of particle are released in cytoplasm
- Retroviral particle contains:
  - two copies of RNA genome
    - template for proviral DNA synthesis
  - specific cellular tRNA
    - primes synthesis by reverse transcriptase
  - reverse transcriptase (RT)
    - performs DNA synthesis
  - integrase
    - carries out DNA integration
- Proviral DNA synthesis is cytoplasmic
  - product is one double stranded DNA molecule
  - RNA template is destroyed during proviral DNA synthesis
- Only one molecule of proviral DNA made per infected cell
  - any mutations made by reverse transcriptase are incorporated during this reaction
  - high mutation rate of HIV-1 can only be achieved by infecting high numbers of cells

Relationship of proviral DNA to genomic RNA

- Proviral DNA is larger than genomic RNA

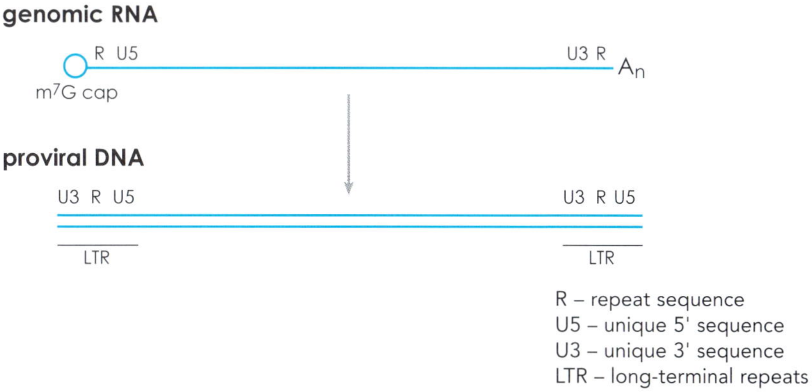

Integration of proviral DNA

- Complex of proviral DNA and integrase migrates to nucleus
- Integrase cleaves proviral DNA, removing two bases at each 3' end
- Integrase uses recessed 3'OH ends of proviral DNA to cleave host DNA
  - no preference for sites on host DNA
- Proviral DNA and host DNA are then joined
  - gaps filled in by host DNA polymerase
- Integration of proviral DNA occurs only in dividing cells
  - exception: HIV-1
    - proviral DNA can be integrated into resting T cells
    - HIV-1 protein Vpr participates in nuclear transport

Consequences of proviral DNA integration

- Cell with integrated proviral DNA is infected for life
- Proviral DNA and thus infection are hidden from immune system
  - medical importance: HIV-1, HTLV-1
- Presence of transcriptional elements in proviral DNA may affect gene expression in infected cell
  - may contribute to cancer
  - veterinary importance: ALV

3.5 Explain in outline the replicative cycle of human T-cell lymphotropic virus 1 (HTLV-1). How is HTLV-1 transmitted and how does it deal with the host immune system?

Human T-cell lymphotropic virus type 1 (HTLV-1)

- Member of retrovirus family
- Enveloped, icosahedral virus with a single stranded RNA genome of positive sense

Replicative cycle

- HTLV-1 only infects helper T cells
  - attaches to unknown protein on surface of these cells
  - virus enters, proviral DNA is synthesised and integrated into host chromosome
- Very little viral RNA produced from integrated HTLV-1 proviral DNA
  - unclear what events bring about increased synthesis of HTLV-1 RNA
- One HTLV-1 protein, called Tax, important for viral infection and propagation
  - Tax causes:
    - increased synthesis of HTLV-1 RNA
    - proliferation of infected cells
    - proliferation of neighbouring uninfected cells
  - consequently, Tax increases:
    - number of HTLV-1 infected cells
    - number of actively dividing helper T cells
    - chances of infection by newly produced virus particles

Transmission

- Almost always as infected cell
  - rarely as free virus
- Virus present in infected cells in:
  - blood
  - semen
  - breast milk
- Transmission via:
  - sexual intercourse
  - blood transfusion and intravenous drug abuse
    - but not via blood products
  - mother to child
    - during birth and breast feeding
- Transmission of cells requires that new host does not destroy new cells
  - HTLV-1 infections restricted to in-bred groups
  - presumably share similar major histocompatibility complex class I molecules

Evasion of immune system

- HTLV-1 is not cytolytic
- Proviral DNA integrates into host chromosome
  - virus not visible to immune system
- Very few HTLV-1 virus particles synthesised
  - little chance of activating immune system

3.6 How can human immunodeficiency virus type 1 (HIV-1) be transmitted? How is HIV-1 not transmitted? How does HIV-1 select and enter a target cell? Why can certain people not be infected with HIV-1?

Human immunodeficiency virus type 1 (HIV-1)

- Member of retrovirus family
- Enveloped, icosahedral virus with a single stranded RNA genome of positive sense

HIV-1 transmission

- Virus and infected cells are present in blood (in the cells and plasma), cerebrospinal fluid, semen, female genital secretions, breast milk (virus and infected cells)

4 most important routes of transmission

- Blood transfusion
  - both with whole blood and blood products
    - now much reduced by screening and treatment
- Re-use of contaminated needles
  - e.g. drug abuse
- Sexual transmission
  - vaginal and anal
  - risk of sexual transmission increased by:
    - frequency of sexual contacts and number of partners
    - infection by other sexually transmitted diseases (esp. those causing ulceration (genital herpes, syphilis) or inflammation or discharge (gonorrhea)
- Mother to child
  - during delivery
  - during breast feeding
  - anti-viral compounds reduce very efficiently transmission of HIV-1 from mother to child
  - note: transmission across the placenta is very low
    - influenced by maternal viral load and presence of other infections

HIV-1 is not transmitted by

- Non-sexual contact
- Kissing (HIV-1 found in saliva only in very low amounts)
- Mosquitoes, ticks, sandflies etc.
    - HIV-1 does not replicate in these arthropods

How does HIV-1 select and enter a target cell?

- Attachment and entry of HIV-1 to cells requires presence of two proteins on cell surface:
- CD4 protein
    - found on helper T cells and macrophages
        - permits attachment, not entry
    - interacts with domain on HIV-1 surface glycoprotein (SU or gp120)
        - binding to CD4 induces conformational change in gp120
- Chemokine receptors
    - most important CCR5 and CXCR4
        - also present on helper T cells and macrophages
    - gp120 binds to chemokine receptors only after conformational change induced by binding to CD4
    - conformational change exposes chemokine binding site
    - binding to chemokine receptor causes conformational change in transmembrane protein (TM or gp41)
        - enables fusion of cellular and viral membranes, leading to viral entry

Why can some people not be infected by HIV-1?

- Gene encoding CCR5 receptor produces defective protein
    - termed CCR5 $\Delta 32$ mutation
    - person with two defective alleles is resistant to HIV-1 infection
        - HIV-1 can still attach to $CD4^+$ cells but cannot enter
    - person with one defective allele has partial resistance and slower progression to disease
    - defective allele widespread in Northern European population
    - hypothesis: previously selected for by another human pathogen
        - most likely candidate: smallpox virus

3.7 Name the genes which HIV-1 possesses in addition to *gag, pol* and *env*. What are the functions of the proteins encoded by these additional genes?

Human immunodeficiency virus type 1 (HIV-1)

- Member of retrovirus family

Retrovirus family

- enveloped, icosahedral viruses with single stranded RNA genome of positive sense
- all retroviruses have three genes: *gag, pol, env*

Proviral DNA of simple retroviruses

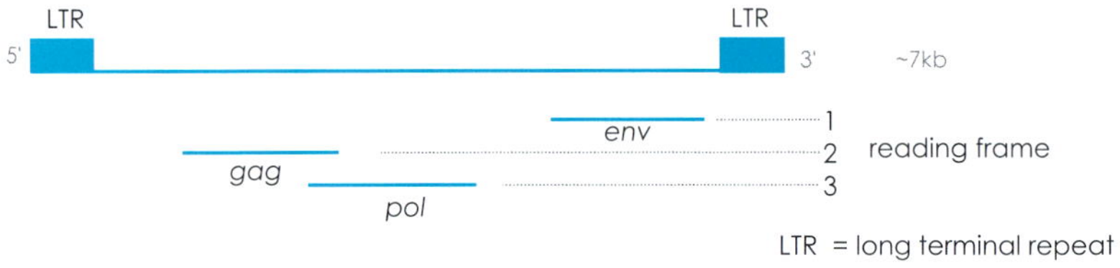

gag encodes	*pol* encodes	*env* encodes
matrix capsid nucleocapsid protease	reverse transcriptase integrase	surface glycoprotein

- HIV-1 has in addition auxiliary genes
  - regulatory genes
    - needed in cell culture
    - *tat, rev*
  - accessory genes
    - not needed in cell culture
    - *vif, vpr, vpu, nef*

Proviral DNA of HIV-1

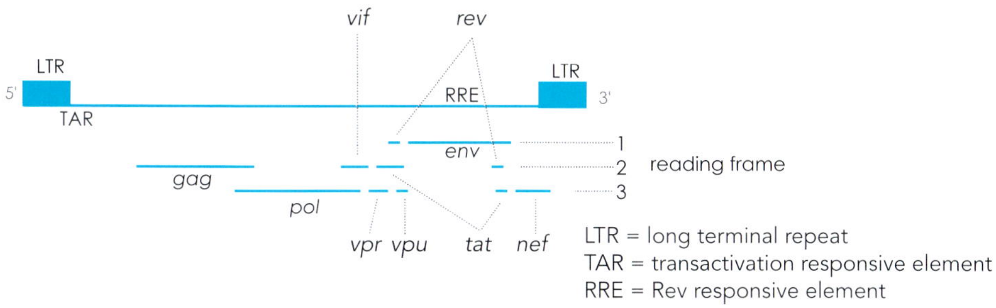

Functions of auxiliary proteins

- Tat (transactivator of transcription)
  - binds to stem-loop structure at 5' end of HIV-1 mRNA
    - called TAR = transactivation responsive element
  - stimulates synthesis of HIV-1 mRNA
- Rev (regulator of viral mRNA production)
  - initially HIV-1 mRNA spliced more than once to generate small mRNAs
    - include those coding for Tat and Rev
  - Tat concentration rises ⇒ viral transcription rises ⇒ Rev concentration rises
  - Rev binds to RRE (Rev responsive element) on HIV-1 mRNA
  - Rev promotes nuclear export of
    - singly-spliced mRNA ⇒ synthesis of Env protein
    - full length mRNA ⇒ synthesis of Gag and Pol proteins
    - full length mRNA is also used as RNA genome in new virus particle
- Nef (negative effector factor)
  - many functions (pleiotropic)
  - can increase or decrease infectivity
- Vpr (viral protein R)
  - causes arrest of cell in G_2-phase
  - facilitates nuclear entry of pre-integration complex of proviral DNA and integrase
- Vpu (viral protein unique)
  - affects virus release
- Vif (viral infectivity factor)
  - increases viral infectivity
  - involved in viral DNA synthesis and viral maturation
  - inactivates host protein called APOBEC3 which can specifically deaminate HIV-1 DNA

3.8 Explain in outline the replicative cycle of HIV-1. How is HIV-1 transmitted and how does it deal with the host immune system?

Human immunodeficiency virus type 1 (HIV-1)

- Member of retrovirus family
- Enveloped, icosahedral virus with a single stranded RNA genome of positive sense

Reproduction of the HIV-1 genome

- Selection of the cells to infect
  - viral envelope glycoprotein targets HIV-1 to $CD4^+$ cells (helper T cells, macrophages)
  - fusion of viral and cell membranes introduces contents of virus particles into cytoplasm
- HIV-1 reverse transcriptase copies two strands of genomic RNA into one molecule of double stranded proviral DNA
  - proviral DNA migrates to nucleus with viral integrase
  - proviral DNA integrated into host chromosome
- Activation of transcription from HIV-1 proviral DNA occurs on activation of helper T cells
  - HIV-1 genomic RNA and mRNAs produced by cellular RNA polymerase II and cellular splicing machinery
- Assembly of virus particle at cell membrane
  - release of HIV-1 by budding from plasma membrane
  - maturation after release by proteolytic cleavages in virus particle

Transmission of HIV-1

- HIV-1 as well as cells infected with HIV-1 are present in:
  - blood, lymph, semen, female genital secretions, cerebrospinal fluid, milk
- Transmission by:
  - blood and blood products – e.g. transfusion
  - re-use of needles – e.g. intravenous drug abuse
  - sex, both anal and vaginal
    - risk enhanced by presence of other sexually transmitted diseases
    - by certain sexual practices
  - mother to infant
    - during birth
    - during breast feeding

How does HIV-1 evade the immune system?

- Initial HIV-1 infection fought by:
  - HIV-1 specific cytotoxic T cells
  - B-cells synthesise large amounts of specific antibodies to HIV-1 which fail to neutralise the virus
    - appearance in blood of specific antibodies to HIV-1 termed seroconversion
  - nevertheless, virus not completely cleared from body
  - still not clear why
- HIV-1 which survives initial attack evades the immune system by:
  - infecting resting helper T cells
    - no expression of HIV-1 proviral DNA
  - actively killing helper T cells
  - infecting and disrupting macrophage function
  - rapidly mutating so that circulating antibodies are directed against a HIV-1 strain which no longer exists
  - having surface glycoprotein which hides conserved areas for receptor binding
    - receptor binding areas are recessed or only exposed after conformational changes
  - evading cytotoxic T cell killing

3.9 What happens during an HIV-1 infection to the
- helper T cells
- viral load
- HIV-1 specific cytotoxic T cells and antibodies

Human T-cell lymphotropic virus (HTLV-1) is also a human retrovirus. Why does the outcome of an infection with HTLV-1 differ from one with HIV-1?

Human immunodeficiency virus type 1 (HIV-1)

- Member of retrovirus family
- Enveloped, icosahedral virus with a single stranded RNA genome of positive sense
- Infects and kills $CD4^+$ cells such as helper T cells and macrophages

What happens to the total number of helper T cells during HIV-1 infection?

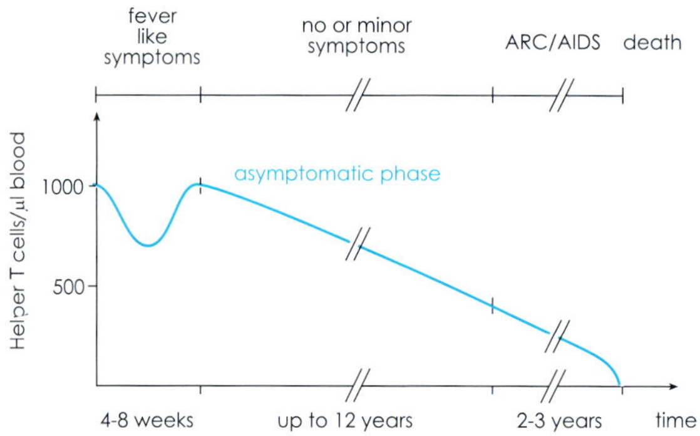

- Characteristic of HIV-1 infection is the steady depletion of helper T cells
- Infected helper T cells are destroyed by:
  - cytolytic nature of HIV-1
  - cytotoxic T cell mediated killing
  - other mechanisms such as immune activation have been proposed
- Stages of HIV-1 infection are defined by the helper T cell count
  - between 500 and 200 helper T cells/µl blood, ARC (AIDS related complex)
  - less than 200 helper T cells/µl blood, AIDS (acquired immunodeficiency syndrome)
- Severely weakened immune system leads to death from infections with normally harmless bacterial, fungal or viral pathogens

What happens to viral load during HIV-1 infection?

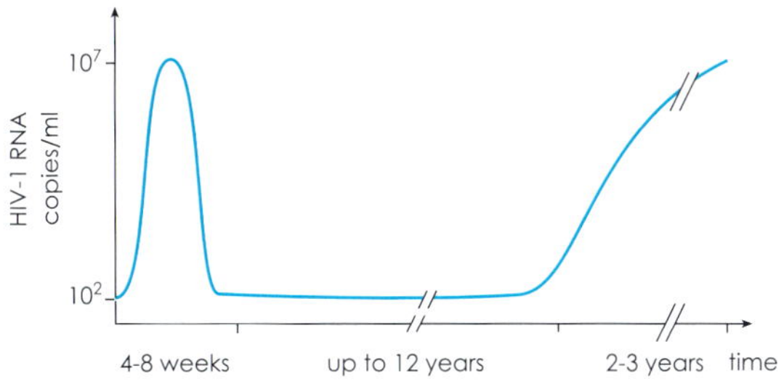

- Viral load is defined as number of HIV-1 RNA copies/ml blood
  - viral load one year after infection termed "viral set-point"
  - indicates how fast progression to AIDS will be
    - high viral set-point indicates a rapid progression
- Initial viral load reduced mostly by cytotoxic T cells
  - but HIV-1 is never completely cleared
    - reason why unknown
- HIV-1 is produced during asymptomatic phase, especially in lymph nodes
  - HIV-1 specific cytotoxic T cells kill infected cells
  - cytotoxic T cells lose effectivity when helper T cells are depleted
    - viral load in blood rises

What happens to HIV-1 specific cytotoxic T cells and antibodies during HIV-1 infection?

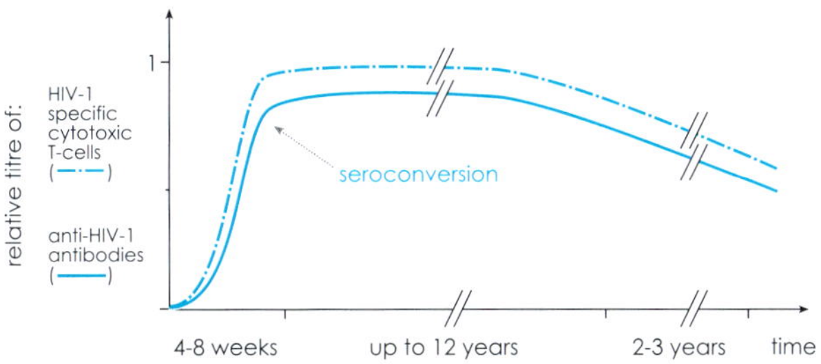

- High antibody titre rapidly reached and maintained, but antibodies fail to neutralise virus

Why does the outcome of an HTLV-1 infection differ from an HIV-1 infection?

- HTLV-1 also infects helper T cells but not macrophages
- HTLV-1 is not cytolytic and does not kill helper T cells
- HTLV-1 produces very little virus
  - very little cytotoxic T cell killing of helper T cells
  - HTLV-1 does not deplete helper T cells

3.10 To understand the transmission of viruses, it is important to know which body fluids may contain viruses. Name five body fluids which may contain viruses and give two examples of a virus of medical importance which can be found in each of them. How did the transmission rates for HIV-1 and hepatitis C virus differ in South Africa between 1991 and 2000? Give reasons for the difference.

Medically important viruses in body fluids

BODY FLUID	VIRUS
blood	hepatitis B virus hepatitis C virus (HCV)
semen	human immunodeficiency virus type 1 (HIV-1) human T-cell lymphotropic virus type 1
saliva	Epstein-Barr virus rabies virus
respiratory fluid (e.g. droplets)	measles virus influenza A virus
faeces	human rotavirus human adenovirus

- Note: The presence of a virus in one fluid does not exclude it from being present in others
    - e.g. HIV-1 is found in:
        - blood
        - semen
        - female genital secretions
        - lymph
        - cerebrospinal fluid
        - milk
    - but HIV-1 is not found in saliva in significant amounts

Comparison of transmission of HIV-1 and HCV

- Human immunodeficiency virus type 1 (HIV-1)
  - enveloped, icosahedral virus with a single stranded RNA genome of positive sense
  - major routes of transmission
    - anal and vaginal intercourse
    - mother to child
    - reuse of hypodermic needles
    - transfusion of contaminated blood and blood products
- Hepatitis C virus (HCV)
  - enveloped, icosahedral virus with a single stranded RNA genome of positive sense
  - major routes of transmission
    - reuse of hypodermic needles
    - mother to child
    - transfusion of contaminated blood and blood products

Transmission rates for HIV-1 and HCV in South Africa 1990-2000

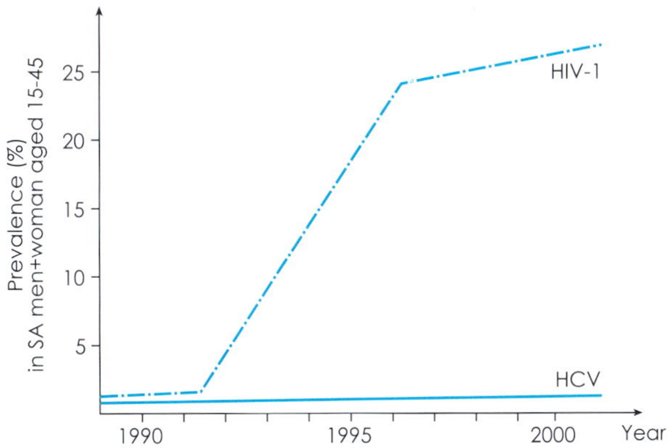

- Reasons for the difference
  - HCV is not transmitted sexually
  - HIV-1 is transmitted sexually

4 DNA viruses

4.1 Sketch the herpes simplex virus type 1 particle. Outline the steps needed for a productive infection by this virus.

Stages in a viral replicative cycle

- Attachment
- Entry
- Uncoating
- Expression of genetic material
- Replication of genetic material
- Assembly
- Maturation
- Release

Herpes simplex virus type 1 (HSV-1)

- Member of herpesvirus family
- Enveloped, icosahedral virus with a linear double stranded DNA genome

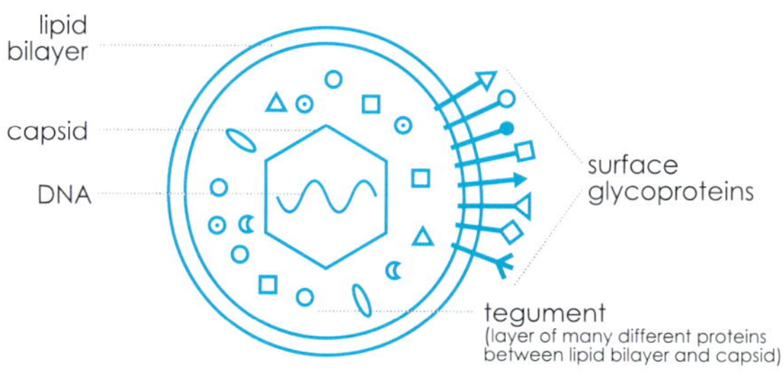

- Infects:
  - epithelial cells productively and lytically
  - nerve cells latently

Replicative cycle of HSV-1 in epithelial cells

- Attachment
  - a HSV-1 surface glycoprotein binds to cellular glycosaminoglycan (e.g. heparan sulphate)
- Entry
  - second HSV-1 surface glycoprotein binds to second cellular molecule
    - induces fusion of viral and cellular membranes
- Uncoating
  - tegument proteins are released into cytoplasm, some migrate into nucleus
    - tegument proteins affect both viral and cellular processes
  - capsid migrates to nucleus
  - linear DNA is released into nucleus where it circularises
- Expression of genetic material
  - transcription of viral DNA by cellular RNA polymerase II
  - 3 classes of transcripts: immediate early, early and late
- Replication of genetic material
  - carried out mostly by proteins encoded by early genes
  - DNA synthesised as long oligomers (termed concatamers) from circularised genomic DNA molecule
- Assembly and maturation
  - involves late proteins
  - DNA is cleaved into monomers and packaged into capsids in the nucleus
    - some tegument proteins bind to capsid in nucleus
  - capsid-tegument complex passes through nuclear membrane into cytoplasm where more tegument proteins are added
  - viral envelope and glycoproteins are collected during passage through Golgi apparatus
- Release
  - mature HSV-1 released from cell by exocytosis

4.2 Herpes simplex virus type 1 and type 2 (HSV-1 and HSV-2) cause labial (cold sores) and genital herpes, respectively. Explain:

- How these viruses cause latent, life-long infections?
- What causes these viruses to leave the latent state and become active?
- Why do these viruses not spread throughout the body?
- Under which circumstances can these viruses cause a systemic infection?

Herpes simplex virus types 1 and 2 (HSV-1 and HSV-2)

- Members of herpesvirus family
- Enveloped, icosahedral viruses with a linear double stranded DNA genome

How do herpes simplex viruses cause latent, life-long infections?

- HSV-1 and HSV-2 can:
  - productively infect epithelial cells
  - latently infect neuronal cells
- In epithelial cells
  - full programme of viral gene expression is carried out
  - large amounts of new virus are made, cells are destroyed
- In neuronal cells
  - almost no gene expression, hardly any HSV proteins produced
  - one characteristic RNA molecule detectable
    - termed LAT, latency associated transcript
    - function unknown
- Virus is invisible to innate and adaptive immune systems

What causes these viruses to leave the latent state and become active?

- Stress, physical or mental
- Hormonal changes (e.g. during menstrual cycle)
- UV-light
- Not clear whether all reactivation factors for HSV-2 are known
- How reactivation takes place in molecular terms is unknown

Why do these viruses not spread throughout the body?

- Two reasons for the lack of spread:
    - first, in immunocompetent patient, strong cytotoxic T cell and natural killer cell response limit infection
    - second, cellular attachment molecule for HSV-1 and HSV-2 is heparan sulphate
        - very abundant molecule, part of extracellular matrix
        - virus is efficiently absorbed to heparan sulphate before it can spread

Under which circumstances can these viruses cause a systemic infection?

- Require an immunosuppressed host and a whole body contact with HSV-1 or HSV-2
    - conditions fulfilled during delivery of child from mother with genital herpes (HSV-2)
    - if mother is experiencing outbreak of genital herpes at the time of delivery, high risk of infection of new-born
    - new-born child has underdeveloped immune system
        - such infections at birth often fatal

4.3 Explain in outline the replicative cycle of human papillomavirus. How is this virus transmitted and how does it deal with the host immune system?

Human papillomavirus (HPV)

- Member of papillomavirus family
- Non-enveloped, icosahedral virus with a circular double stranded DNA genome

The replicative cycle of human papillomavirus

- Attachment and entry
  - HPV initially infects basal stem cells under skin and mucosal surfaces
    - HPV binds to unknown cellular molecule to enter the cell
  - DNA delivered to nucleus, perhaps by uncoating at the nuclear membrane
    - HPV DNA remains episomal (i.e. not integrated into the host genome)
- Gene expression and replication of genetic material
  - in basal stem cells
    - only some HPV genes expressed
    - no production of virus particles
  - when stem cells differentiate to keratinocytes
    - all genes expressed
    - viral DNA replicated by host cell machinery
    - virus particles produced
- Assembly and release
  - assembly and release of new virus unclear
  - assembly thought to occur in nucleus
    - DNA bound by cellular histones before packaging
  - release occurs without cell lysis

Transmission of human papillomavirus

- Physical contact
  - sexual contact
  - contact sports
  - children's games
- During birth
  - from infected mother to child

Human papillomavirus and immune system

- Innate immunity has difficulties to recognise HPV infection
  - HPV is not cytolytic
  - HPV has no envelope
  - HPV does not express proteins which insert in the cell membrane of infected cells
  - HPV expresses its genes from only one DNA strand
    - no complementary mRNAs
    - no double stranded RNA intermediate
- No innate immunity, no adaptive response
- HPV infection may persist undetected for several months
  - remains undetected even when HPV infection leads to formation of warts
    - however, warts may be super-infected by bacteria, leading to inflammation
    - immune system then detects both bacteria and papillomavirus
- Once detected, infection cleared by cytotoxic T cells
  - explains sudden and simultaneous disappearance of warts

4.4 Which diseases are caused by adenoviruses? Sketch the basic genome structure of a human adenovirus. Why is the adenoviral DNA bound to a viral protein at its 5' end? Which strategy does adenovirus use to express several proteins from one transcriptional unit?

Adenovirus family

- Non-enveloped, icosahedral viruses with a double stranded linear DNA genome

Which diseases are caused by adenoviruses?

- Diseases caused by adenoviruses include
  - infant diarrhoea
  - conjunctivitis
  - respiratory diseases (including common cold)
- Many infections are however subclinical

Genome structure of human adenovirus

TP, terminal protein
R, repeat sequence
r, complement to repeat sequence
ori, origin of replication

Why is the adenoviral genome bound to a viral protein at its 5' end?

- TP, terminal protein
  - originates from initiation of DNA synthesis on the adenovirus genome
- Mechanism of DNA synthesis:
  - adenoviral DNA polymerase cannot initiate new DNA chain
  - requires protein primer called preterminal protein (pTP)
  - DNA polymerase complexed with pTP binds to ori
  - first nucleotide covalently linked to pTP
  - DNA polymerase can then elongate new chain
- pTP cleaved to TP by adenoviral protease during maturation within virus particle

Which strategy does adenovirus use to express several proteins from one transcriptional unit?

- Adenovirus genetic information expressed as transcriptional units (TU)
  - transcriptional unit = DNA region from start to finish of RNA synthesis
  - may encode several proteins
  - reduces number of transcription signals required
- In adenoviruses, primary RNA transcripts are generally spliced to smaller mRNAs
  - e.g. E1A mRNAs are generated from one TU
  - differential splicing produces 5 mRNAs from primary transcript
  - encoded proteins have different functions
    - some modify cell physiology
    - some promote transcription from other TUs
- Use of TUs and splicing enables adenoviruses to conserve genome space
  - genome space limited by capsid size

4.5 What are the functions of simian virus 40 (SV40) large T antigen?

Simian virus 40 (SV40)

- Member of the polyomavirus family
- Non-enveloped, icosahedral virus with a circular double stranded DNA genome
- Genome possesses two transcriptional units: early and late
  - early transcriptional unit encodes small T and large T antigens
  - late transcriptional unit encodes three capsid proteins
  - differential splicing is used to generate more than one protein per transcriptional unit

Large T antigen

- Protein of about 700 amino acids
- Functions as hexamer
- Some activities require phosphorylation at residue 124

Functions of large T antigen

- Binds to ori region of SV40 genome
    - represses transcription from early transcriptional unit
        - i.e. repressses its own transcription
    - activates transcription from late transcriptional unit
- Binds to cellular DNA polymerase in species-specific manner
    - i.e. binds simian (monkey) enzyme but not murine (mouse) enzyme
- Acts as ATP dependent DNA helicase
    - together with host DNA polymerase replicates SV40 genome
- Binds to cellular protein Rb
    - Rb normally represses synthesis of proteins required for DNA synthesis (S-phase proteins)
    - binding of large T antigen inactivates Rb
        - large T antigen actively dismantles complexes of Rb with its binding partners
        - S-phase proteins can be synthesised
        - replication of SV40 genome can proceed
- Binds to cellular protein p53
    - p53 normally senses unscheduled DNA synthesis brought about by Rb inactivation
        - p53 actively leads to cell cycle arrest and apoptosis
    - binding of large T antigen inactivates p53
        - allows unscheduled DNA synthesis to take place
        - evades cellular defence mechanisms

Summary

- Entire life-cycle of SV40 controlled by single, versatile protein

4.6 Describe in outline how two DNA viruses replicate their genetic material. One of your chosen viruses should use a cellular polymerase whilst the other should use a viral DNA polymerase.

Simian virus 40 (SV40)

- Member of polyomavirus family
- Non-enveloped, icosahedral virus with a double stranded circular DNA genome
- Uses cellular DNA polymerase
- Genome structure:

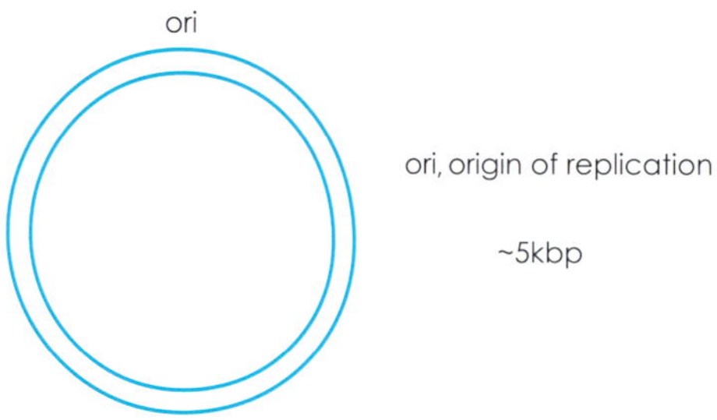

- Replication of genetic material
  - SV40 encodes protein called large T (Tumour) antigen
    - large T antigen binds to ori and unwinds SV40 DNA
    - large T antigen binds to cellular proteins such as single stranded DNA binding protein and DNA polymerase α
    - DNA polymerase α initiates replication of DNA genome by synthesising RNA primers
    - DNA polymerase δ extends RNA primers into longer chains
  - replication proceeds bi-directionally from ori

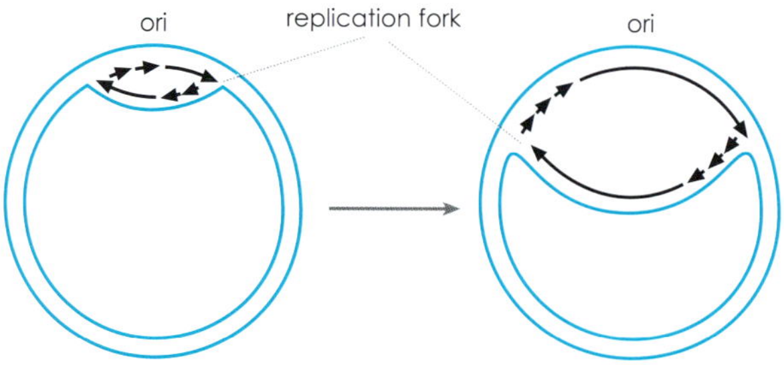

  - separation of DNA molecules and movement of replication forks require topoisomerases

Human adenovirus

- Member of adenovirus family
- Non-enveloped, icosahedral virus with a double stranded linear DNA genome
- Encodes its own DNA polymerase
- Genome structure:

TP, terminal protein
R, repeat sequence
r, complement to repeat sequence
ori, origin of replication

- Replication of genetic material
    - viral proteins: DNA polymerase, preTerminal protein (pTP), single stranded DNA binding protein (ssDbp)
    - cellular proteins: transcription factors Nf-1 and Oct-1, topoisomerase
    - replication initiated by binding of pTP and polymerase to ori
        - transcription factors bind to sites on ori and recruit pTP and polymerase
        - new chain initiated by linking first nucleotide to pTP

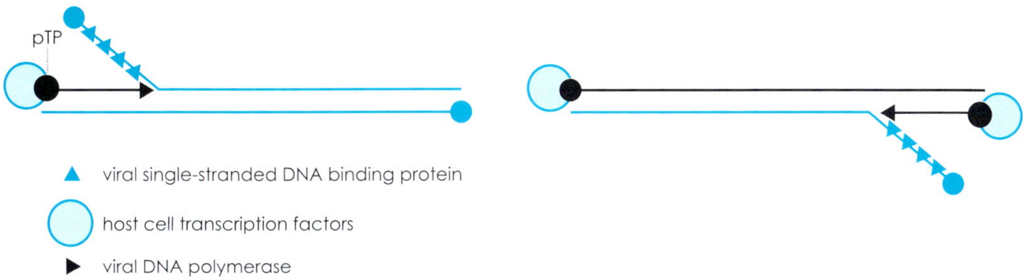

        - non-copied strand displaced
        - DNA synthesis in the opposite direction can be carried out similarly
        - pTP cleaved by viral protease to TP during maturation
    - displaced strands are single stranded but 5' end and 3' end are complementary
        - form hairpin structures

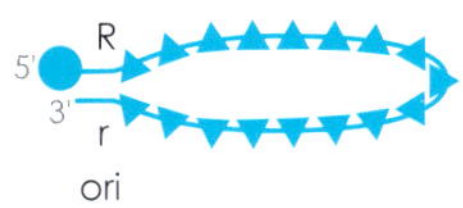

        - presence of double stranded ori allows this molecule to be copied similarly

4.7 Give four examples of the viruses which employ RNA splicing to generate mRNAs encoding different proteins.

RNA splicing

- Conversion of primary RNA transcript into one or more smaller RNAs
    - carried out in nucleus of the host cell

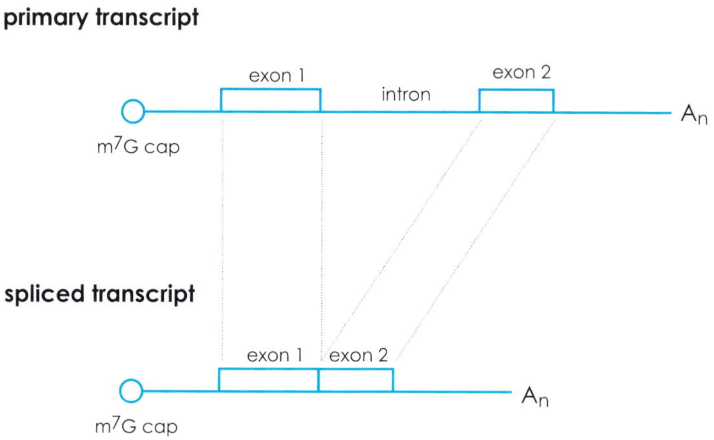

Simian virus 40 (SV40)

- Member of polyomavirus family
- Non-enveloped, icosahedral virus with a circular double stranded DNA genome
- RNAs from both early and late transcriptional units (TUs) are spliced
    - e.g. two mRNAs produced from early transcriptional unit by splicing
        - encode large T antigen and small T antigen
        - N-terminal region of both proteins identical

Human adenovirus

- Member of adenovirus family
- Non-enveloped, icosahedral virus with a linear double stranded DNA genome
- Primary transcript from E1A transcriptional unit is spliced into five different mRNAs
    - all five proteins produced have the same N-terminus
    - larger proteins are transcriptional activators
    - smaller proteins interact with host cell

Avian leukosis virus (ALV)

- Member of retrovirus family
- Enveloped, icosahedral virus with a single stranded RNA genome of positive sense
- Unspliced primary transcript from ALV provirus is used as:
    - mRNA for proteins from *gag* and *pol* genes
    - genomic RNA for packaging into new virus particles
- Primary transcript can also be spliced to generate mRNA encoding for protein from *env* gene

Influenza A virus

- Member of orthomyxovirus family
- Enveloped, helical virus with eight single stranded RNAs of negative sense
- Although influenza is an RNA virus, its gene expression and replication are nuclear
    - influenza virus employs host cell transcription and splicing machinery
- Splicing takes place on mRNAs generated from RNA segments 7 and 8
    - the mRNAs can either be transported from nucleus without splicing
        - e.g. unspliced mRNA from segment 7 encodes matrix protein M1
    - or mRNAs can be spliced to smaller mRNAs
        - e.g. spliced mRNA from segment 7 encodes matrix protein M2

4.8 Human papillomavirus and herpes simplex virus 1 can both cause long-term infections. Explain how these viruses cause long-term infections and point out the important differences.

Human papillomavirus (HPV)

- Member of papillomavirus family
- Non-enveloped, icosahedral virus with a circular double stranded DNA genome

Strategy of the human papillomavirus

- Avoid detection by immune system
- Avoid innate immunity through:
  - lack of envelope in the particle
    - no detection by macrophages
  - no viral proteins inserted in membrane of infected cell
    - no detection by natural killer cells
  - producing all RNAs from one strand of the genome
    - no double stranded (ds) RNA as replicative intermediate
    - no interferon response
- Absence of innate immune response means no adaptive response
- HPV infects basal stem cells below skin and mucosal surfaces
  - but no virus particles produced in these cells
  - reduces chances of detection by immune system
- Particles only produced in keratinocytes
- HPV strategy is not for life, at some point (may take 1 to 2 years) immune system senses danger
  - immune system reacts to other infections in same area and detects HPV "by accident"
    - e.g. superinfection by bacteria of warts caused by HPV may lead to lysis of keratinocytes and thus release of HPV components
  - HPV infection usually then cleared by strong immune response, especially by cytotoxic T cells

Herpes simplex virus type 1 (HSV-1)

- Member of herpesvirus family
- Enveloped, icosahedral virus with a linear double stranded DNA genome

Strategy of the herpes simplex virus type 1 (HSV-1)

- Interfere with and hide from immune system
- HSV-1 has surface glycoproteins which
  - inactivate complement C3b
  - inactivate antibodies
- HSV-1 infection reduces expression of major histocompatibility complex class I molecules
  - reduces efficiency of adaptive immunity
  - however, this reduction can be sensed by natural killer cells
- HSV-1 productively infects epithelial cells
  - infection cleared by adaptive immunity
- Nearby neurons also infected
  - neurons infected latently
    - no viral replication, no expression of viral proteins
    - only latency associated RNA transcripts (LATs) expressed
  - virus is effectively hidden from immune system
- Infected neurons never cleared
- HSV-1 reactivated from latency
  - by stress, UV light, hormonal changes
- Virus released from neurons without their lysis
  - leads to productive infection of neighbouring epithelial cells
  - infection of epithelial cells cleared by immune system
- HSV-1 strategy is for life

4.9 Classify Epstein-Barr virus (EBV). Which cells does EBV infect? Name three diseases which are associated with EBV.

Epstein-Barr virus (EBV)

- Member of herpesvirus family
- Enveloped, icosahedral virus with a linear double stranded DNA genome

Which cells does EBV infect?

- EBV infects epithelial cells of the throat and B lymphocytes
  - productive replication only in cells of the throat
  - replication of EBV in B lymphocytes causes them to divide
- Most infected cells cleared by immune system
- However some infected B lymphocytes escape immune surveillance
  - not recognised by cytotoxic T cell response
  - allows latent infection of memory B cells
    - only one EBV protein expressed
- EBV can be reactivated from memory B cells
  - signals responsible and mechanism are unclear

Diseases associated with EBV

- Infectious mononucleosis (glandular fever or "kissing disease")
  - EBV infection of B cells disrupts immune system
  - this disease only found when an EBV infection occurs in adolescence or early adulthood
  - EBV infection in childhood is usually subclinical
    - childhood infections very frequent in non-industrialised countries
  - in industrialised countries, stricter hygiene prevents exposure to the virus until adolescence
    - often first encountered during kissing
    - explains incidence of infectious mononucleosis in industrialised countries

- Burkitt's lymphoma
  - cancer of B cells
    - affects children in New Guinea and certain parts of South America and Africa
  - risk factors are:
    - EBV infection in early life
    - repeated infection with *Plasmodium falciparum* (causative agent of malaria)
    - life below 1000 metres in high rainfall area
  - interactions of risk factors unclear
    - *P. falciparum* may cause immunosuppression and/or incorrect T cell response
  - nevertheless, one common feature of almost all Burkitt lymphoma cells is presence of specific chromosome translocation that places cellular *myc* gene under the control of immunoglobulin enhancer
- Nasopharyngeal cancer
  - one of the most common cancers of Chinese males
  - latent EBV present in all populations in the world
    - but incidence of nasopharyngeal cancer much higher amongst Chinese males
  - why EBV infection in this genetic group frequently leads to nasopharyngeal cancer is still unclear

4.10 Describe the functions of the proteins Rb and p53. Which viral proteins can affect the function of the host proteins Rb and p53? Which advantages do these interactions have for the viruses?

Functions of host proteins Rb and p53

- Rb and p53 are anti-oncogenic proteins
- Rb:
  - prevents expression of genes required for DNA synthesis phase of cell cycle
  - binds and inactivates specific transcription complex E2f
  - in healthy cell, Rb degraded at correct time to permit synthesis of mRNA for proteins required for DNA synthesis
- p53:
  - prevents unscheduled DNA synthesis
  - recognises damaged DNA and stops DNA synthesis until DNA is repaired
  - activation of p53 can have two outcomes
    - arrest of cell cycle before the start of DNA synthesis phase
    - induction of programmed cell death (apoptosis)

- Viral proteins that interact with Rb and p53

VIRUS	VIRAL PROTEIN	INTERACTS WITH
SV40	large T antigen	Rb, p53
human papillomavirus	E6	p53
	E7	Rb
adenovirus	E1A	Rb
	E1B, E4	p53

Which advantages do these interactions have for the viruses?

- Binding to Rb
  - DNA viruses depend on host cell to replicate their genomes
  - binding of viral proteins to Rb allows E2f to become active
  - E2f stimulates S-phase gene expression
    - viral DNA genome can be replicated
- Rb has relatives which maintain cells in quiescent state
  - such cells do not re-enter the cell cycle
  - viral proteins bind to Rb relatives and inactivate them
    - enable quiescent cells to re-enter cell cycle
- Binding to p53
  - unscheduled DNA synthesis activates p53
    - activated p53 causes cell cycle arrest or apoptosis
  - binding of viral proteins prevents p53 from performing this task
  - prevents cell from destroying itself and the virus with it

Medical importance

- Human papillomavirus (HPV)
  - risk factor for cervical carcinoma
  - all genotypes encode E6 and E7 proteins
  - however, only certain genotypes (e.g. HPV 16 and 18) are oncogenic
    - E6 and E7 proteins of these genotypes must be different

4.11 Human papillomaviruses can cause skin warts. Explain the mechanism behind this. How can certain human papillomaviruses be a risk factor for cancer?

Human papillomaviruses (HPV)

- Member of papillomavirus family
- Non-enveloped, icosahedral virus with a circular double stranded DNA genome
- Around 80 to 90 genotypes (no serotypes, no serology)
  - some types infect skin
  - others infect mucosal surfaces (e.g. genital tract, respiratory tract)

How do human papillomaviruses cause skin warts?

- Transmission by physical contact
  - virus enters via tears or breaks in skin
- Virus infects basal stem cells below surface of skin
  - basal stem cells proliferate, replicating papillomavirus DNA
  - viral DNA remains separate from chromosomal DNA (i.e. episomal)
    - but no virus produced in these cells
- Stem cells may differentiate to keratinocytes (epithelial cells)
  - keratinocytes do not proliferate
    - but virus particles are produced in these cells
- Papillomavirus produces two proteins E6 and E7
  - E7 allows cellular proteins required for DNA synthesis to be made
  - E6 prevents apoptosis of infected cells normally induced by unscheduled DNA synthesis
- Synthesis of E6 and E7 proteins causes keratinocytes to proliferate
  - causes thickening of the skin
  - warts only generated when extra blood vessels synthesised
    - supply keratinocytes in wart with nutrients
- How the virus causes the synthesis of new blood vessels is not known

How can certain human papillomaviruses be a risk factor for cancer?

- Certain human papillomaviruses which cause genital warts are a risk factor for cervical cancer
- Fate of genital papillomavirus infections
  - most cleared rapidly, no viral genetic material left
  - some persist for year or so, then cleared
  - very minor percent not cleared, viral DNA persists, often integrated into chromosomal DNA
- Persistent infections with high risk genotypes (e.g. genotypes 16 and 18) may develop over many years to cervical cancer
  - DNA from high risk types detected in 99% of all cervical cancers
- What is the mechanism?
  - integration of viral DNA leads to uncontrolled expression of E6 and E7 proteins
  - properties of E6 and E7 proteins from high risk genotypes may be different from those of low risk genotypes
    - perhaps more effective at interfering with cellular proteins

5 Viruses and cancer

5.1 How can viruses be risk factors for cancer? Give examples of two DNA and two RNA viruses that are risk factors for cancer.

Cancer (neoplastic transformation)

- Uncontrolled cell growth
- Two basic mechanisms leading to uncontrolled cell growth
  - presence of (a) factor(s) which accelerate(s) cell cycle
    - e.g. receptor for hormone or growth factor
    - e.g. transcription factor
  - loss of factors which prevent unscheduled DNA synthesis
    - e.g. deletion of or mutation in Rb protein
    - e.g. deletion of or mutation in p53 protein

How can viruses be risk factors for cancer?

- Viruses can encode proteins which accelerate the cell cycle
  - e.g. growth factors, growth factor receptors, transcription factors
    - such viral proteins are often close relatives of cellular proteins
    - but viral proteins lack control elements found in cellular proteins
- Viruses can encode proteins which release the brakes on the cell cycle
  - e.g. proteins which bind and inactivate Rb
    - found in many DNA viruses
  - proteins move cell into S-phase of the cell cycle to allow replication of viral genome
- Viruses encode proteins which transactivate transcription of cellular genes
- Viral transcriptional elements can increase expression of cellular oncogenes

Two DNA viruses which can cause cancer

- Epstein-Barr virus (EBV)
  - member of herpesvirus family
  - enveloped, icosahedral virus with a linear double stranded DNA genome
  - EBV infection in childhood is a risk factor for Burkitt's lymphoma
    - Burkitt's lymphoma is a cancer of B cells
  - other risk factors for this cancer
    - infection with malaria parasite
    - life below 1000 metres in high rainfall area
  - risk factors limit occurrence of Burkitt's lymphoma to certain parts of South America, Africa and New Guinea

- Hepatitis B virus (HBV)
  - member of hepadnavirus family
  - enveloped, icosahedral virus with a partially incomplete double stranded DNA genome
  - risk factor for hepatocellular cancer
  - how HBV causes hepatocellular cancer is unclear
  - hypothesis:
    - persistent infections lead to increased cell division
    - probability of mistakes in DNA replication increases
    - probability of such mistakes further increased by chemicals reaching the liver for detoxification

Two RNA viruses which can cause cancer

- Hepatitis C virus (HCV)
  - member of flavivirus family
  - enveloped, icosahedral virus with a single stranded RNA genome of positive sense
  - risk factor for hepatocellular cancer
  - how HCV causes hepatocellular cancer is unclear
  - same hypothesis as for HBV
- Human T cell lymphotropic virus type 1 (HTLV-1)
  - member of retrovirus family
  - enveloped, icosahedral virus with a single stranded RNA genome of positive sense
  - risk factor for adult T cell leukaemia/lymphoma
  - HTLV-1 encodes a protein (Tax) which transactivates transcription of certain cellular genes

Medical importance

- All four viruses are important risk factors in human health
- Effective vaccine for HBV
- No vaccines available for EBV, HCV and HTLV-1

5.2 Give three examples of viruses which are risk factors for cancer. Explain why infections with two of these viruses are risk factors for cancer. Which vaccines are available to fight against cancer?

Three viruses which are risk factors for cancer

- Hepatitis B virus (HBV)
  - member of hepadnavirus family
  - enveloped, icosahedral virus with a double stranded, circular, incomplete DNA genome
  - persistent infection with HBV is a risk factor for hepatocellular carcinoma (HCC)
- Human papillomavirus (HPV)
  - member of papillomavirus family
  - non-enveloped, icosahedral virus with a double stranded circular DNA genome
  - persistant infection with certain HPV genotypes is a risk factor for cervical cancer
- Human T cell lymphotropic virus type 1 (HTLV-1)
  - member of retrovirus family
  - enveloped, icosahedral virus with single stranded RNA genome of positive sense
  - persistant infection with HTLV-1 is a risk factor for adult T cell leukaemia/lymphoma (ATLL)

Why are infections with two of these viruses risk factors for cancer?

- Hepatitis B virus (HBV)
  - HBV infects hepatocytes (liver cells) but does not kill them
    - cytotoxic T cells cause liver damage during HBV infection by killing infected cells
    - normally infection is cleared and liver regenerates
  - however, if infection occurs in first years of life
    - high probability of persistent infection
    - permanent killing of infected hepatocytes and liver regeneration
    - permanent DNA synthesis in hepatocytes leads to higher probability of mutations
    - probability of mutations further increased by liver's function of handling and detoxifying harmful chemicals
    - the higher the mutation rate, the higher probability of hepatocellular carcinoma (HCC) development
  - this is a hypothesis, it has not been proven
- Human papillomavirus (HPV)
  - HPV infects basal stem cells under skin and mucosal surfaces
    - certain genotypes cause ano-genital warts
    - most infections cleared rapidly (less than 1 month)
    - remainder after 1 year
  - however, some infections persist
    - integration of viral DNA into host chromosome allows high expression rate of HPV proteins
    - expression of proteins from certain high-risk HPV genotypes (HPV-16 and HPV-18) predisposes host to cervical cancer

Which vaccines are available to fight cancer?

- Vaccines are available against:
  - hepatitis B virus (HBV)
    - recombinant subunit vaccine
  - human papillomavirus (HPV)
    - recombinant subunit vaccine
- Vaccines are not available against:
  - human T cell lymphotropic virus type 1 (HTLV-1)
  - hepatitis C virus (HCV)
  - Epstein-Barr virus (EBV)

5.3 How can retroviruses be risk factors for cancer?

Retrovirus family

- Enveloped, icosahedral viruses with a single stranded RNA genome of positive sense
- RNA genome replicated through a DNA intermediate (proviral DNA)

First step in retroviral oncogenesis

- Integration of proviral DNA into host chromosome

Subsequent steps in retroviral oncogenesis

- All retroviral genomes contain elements to promote RNA synthesis
    - if proviral DNA integrates near cellular oncogene, expression of cellular oncogene may be increased
    - probability of cellular transformation increased
    - e.g. avian leukosis virus (ALV)
        - important pathogen of chickens
- Some retroviral genomes encode oncogenes
    - e.g. Rous sarcoma virus (RSV)
        - encodes v-Src, a tyrosine kinase
        - v-Src very similar to cellular counterpart c-Src
        - but v-Src lacks control region and is therefore constantly active
        - permanent expression accelerates cell growth
        - probability of cellular transformation increased
- Some retroviruses may encode proteins which transactivate transcription of cellular genes
    - e.g. human T-cell lymphotropic virus type 1 (HTLV-1)
        - encodes a protein called Tax
        - Tax binds to viral transcription elements to activate transcription
        - also activates transcription of host genes encoding e.g. interleukin-2 and interleukin-2 receptor
        - leads to T-cell proliferation
        - probability of cellular transformation increased

Medical importance

- Adult T-cell leukaemia/lymphoma (ATLL)
  - develops in about 1% of HTLV-1 infected people
  - long latency, about 30 to 50 years after initial infection

5.4 Give two examples of vaccines against oncogenic viruses Which parts of the virus particles do the vaccines contain? Why are the vaccines safe?

Two examples of vaccines against oncogenic viruses

- Hepatitis B virus
  - member of hepadnavirus family
  - enveloped, icosahedral virus with an incomplete circular double stranded DNA genome
  - risk factor for hepatocellular cancer
  - recombinant vaccine used routinely
  - preventing infection with hepatitis B virus prevents hepatocellular cancer
- Human papillomavirus
  - member of papillomavirus family
  - non-enveloped, icosahedral DNA virus with a circular double stranded DNA genome
  - risk factor for cervical cancer
  - recombinant vaccine available
  - preventing infection with human papillomavirus prevents cervical cancer

Which parts of the virus particles do the vaccines contain?

- Hepatitis B virus vaccine contains surface antigen (HBsAg)
  - HBsAg produced in yeast using recombinant DNA technology
  - HBsAg self-assembles to virus-related particles
  - virus-related particles induce immunity against virus
- Human papillomavirus vaccine contains capsid protein VP1
  - produced in baculovirus using recombinant technology
  - VP1 self-assembles to virus-like particles
    - no other capsid protein required
  - virus-like particles induce immunity against papillomavirus
  - recombinant vaccine contains epitopes to four high risk papillomavirus genotypes - vaccine "quadrivalent"

Why are the vaccines safe?

- Only one viral component, no viral genetic material
  - no reversion
  - no incomplete inactivation
  - no recombination with related viruses
- No other viruses present

6 Viruses and the immune system

6.1 What are the four components of innate immunity? How can innate immunity deal with viruses inside cells?

Innate immunity

- Senses invaders without prior information
- Recognises invaders through surface structures or molecular patterns produced during their replication
- Prevents infection and thus disease

The four components of innate immunity

- Professional phagocytes
  - ingest and destroy invaders (phagocytosis)
  - macrophages
    - present in tissues
    - young macrophages recruited from blood
    - recognise unusual cell death caused by viral replication
    - recognise invaders by presence of certain lipids or sugars not normally found in host
    - recognition activates macrophages and induces release of tumour necrosis factor α (TNFα)
    - TNFα activates natural killer cells
  - neutrophils
    - present in blood
    - recruited into tissues by signals from macrophages and presence of certain complement proteins
- Natural killer cells
  - kill infected cells without further instruction if:
    - viral molecules detected on surface
    - host molecules (e.g. major histocompatibility complex class I) missing
  - activation of natural killer cells causes release of interferon γ which activates macrophages

- Complement proteins (e.g. complement protein C3)
  - C3 undergoes constant cleavage into products C3a and C3b
    - C3b can bind to certain molecular structures only found on invaders
  - binding of C3b to an invader
    - increases cleavage efficiency of protein C3
    - initiates assembly of complex which can lyse enveloped viruses
    - induces uptake of coated invaders by macrophages and increases activation state of macrophages
  - rise in C3a concentration recruits neutrophils
- Cytokines
  - e.g. interferons (IFN) α, β and γ, tumour necrosis factor α (TNFα)
  - IFNα and IFNβ
    - produced by almost all cells in response to viral infection
    - cause infected cells to die
    - prepare neighbouring non-infected cells for cell death if they become infected
  - IFNγ
    - produced by natural killer cells and helper T cells
    - activates macrophages
  - TNFα
    - produced by macrophages
    - activates natural killer cells
  - IFNγ and TNFα
    - coordinate innate immune response
    - initiate adaptive immune response (T and B cells) if required

Innate immunity deals with intracellular viruses using

- IFNα, IFNβ and natural killer cells

6.2 How can the innate immune system recognise viruses or signs of viral replication? How does the innate immune system instruct the adaptive system?

How can the innate immune system recognise viruses or signs of viral replication?

- Presence of double stranded (ds) RNA
  - dsRNA is not generated in uninfected cells
  - generated during replication of many RNA viruses
    - use dsRNA intermediate to replicate RNA genome
  - generated during replication of many DNA viruses
    - e.g. by complementary RNA molecules transcribed from both strands of DNA genome
  - leads to interferon synthesis
- Presence of unusual secondary or tertiary RNA structures
  - e.g. the internal ribosome entry site (IRES) in picornaviral RNAs
    - higher ordered RNA structure required for initiation of protein synthesis
  - leads to interferon synthesis
- Unusual cell death
  - cells may be lysed by cytotoxic (cytolytic) viruses
    - e.g. herpes simplex virus type 1, influenza virus
  - cells may enter programmed cell death via interferon response
  - sensed by macrophages
- Presence of unusual sugars on virus particle
  - e.g. surface glycoprotein of viral envelope
  - sensed by macrophages, complement proteins C3b, C1q and mannose binding lectin
- Presence of unusual sugars on the surface of infected cell
  - sensed by natural killer cells
- Changes in major histocompatibility complex (MHC) expression
  - MHC class I molecules display contents of the cells to the immune system
    - some viruses reduce expression of MHC molecules
    - e.g. herpes simplex virus type 1
  - sensed by natural killer cells

How does the innate immune system instruct the adaptive system?

- Macrophages and natural killer cells produce cytokines upon recognition of invader
- Cytokines
  - include: tumour necrosis factor α, interleukin 8, interferon γ
  - activate dendritic cells (DCs)
- DCs present in all tissues
  - in inactive state, permanent sampling of tissue fluid
  - upon activation, DCs migrate to lymph nodes
    - at lymph node, DCs present antigens to helper T cells
- Helper T cells activate antigen specific B cells and are involved in activation of cytotoxic T cells
  - cytotoxic T cells travel to infected region
    - rechecked for antigen specificity by macrophage antigen presentation
- The adaptive immunity is only activated when an innate immune response is triggered
- Innate immunity tells adaptive immunity what type of invader is present and where it is invading

6.3 What are interferons? How are interferons induced? How can interferons inhibit viral replication?

What are interferons?

- Interferons (IFNs) are cytokines
- Produced in response to viral infection
- 3 types: IFNα, IFNβ and IFNγ

IFNα and IFNβ

- Produced by almost every cell
- Signal the anti-viral state
- Prepare the infected cell for self-destruction e.g. by apoptosis
- Prepare neighbouring cells for self-destruction when and if they become infected
- Increase expression of major histocompatibility complex class I molecules for antigen presentation to cytotoxic T cells

IFNγ

- Produced by natural killer cells and certain types of T cells
- Activates macrophages to act as antigen presenting cells
- Influences B cells to produce a certain antibody class

Induction of the synthesis of IFNα and IFNβ

- By molecules indicating viral replication
  - e.g. viral structural proteins on the cell surface
  - e.g. double stranded RNA (dsRNA)
  - dsRNA generated:
    - in RNA viruses as an intermediate in genome replication
    - in DNA viruses by transcription of mRNA from both strands
- Viral presence often detected by Toll-like receptors (TLRs)
  - signalling by TLRs leads to induction of IFNα and IFNβ

Induction of synthesis of IFNγ

- Recognition of infected cell by natural killer cells
- Activation of certain helper T cells

How can interferons inhibit viral replication?

- IFNα and IFNβ, only bind to cells with receptor for IFNα and IFNβ
- Receptor binding induces synthesis of several hundred proteins
  - e.g. protein kinase (Pkr)
    - synthesised in inactive state
    - activated by presence of dsRNA
    - activated Pkr inhibits protein synthesis ⇒ viral replication inhibited
    - inhibition of protein synthesis leads to self-destruction of infected cell by autophagy (cells literally "eat themselves up")

Consequences of action of IFNα and IFNβ

- Infected cell generally destroys itself
- Uninfected cell is primed to destroy itself, but does not do so until it senses signs of viral replication

6.4 Summarise innate and adaptive immunity. Compare and contrast their modes of action. In which way do they work together?

Human defences against pathogens

- First line: skin and mucosal surfaces
- Second line: innate immune system
- Third line: adaptive immune system

Innate immunity (II)

- The II response fights pathogens using:
  - professional phagocytes (macrophages and neutrophils)
  - complement system (above all C3 protein)
  - cytokines (e.g. interferons and tumour necrosis factor)
  - natural killer cells
- If the II response is overwhelmed, the adaptive immunity fights the invader

Adaptive immunity (AI)

- The AI response fights pathogens using:
  - cytotoxic T cells
  - antibodies produced by B cells
  - helper T cells
    - control and activate B cells and cytotoxic T cells
- AI has memory
  - memory B and T cells produced after infection
  - remain to protect against reinfection with same pathogen
  - vaccination usually aims to generate at least memory B cells

Modes of action of II and AI

II	AI
rapid	slow
non-specific	specific
does not need to be taught	must be taught
can prevent infection	cannot prevent primary infection
no memory	memory - can protect against reinfection

- II recognises pathogens by:
  - surface structures not found in host
    - e.g. certain carbohydrates and lipids
  - presence of RNA secondary structure or double stranded RNA
- AI generally recognises pathogens using proteins
  - AI instructed using small peptides presented on major histocompatibility complex proteins
  - T and B cells recognising host proteins are eliminated during development

How do II and AI work together?

- If there is no II response to an invader, there cannot be an AI response
- II instructs AI which type of response is required
  - macrophages and natural killer cells produce cytokines
    - above all interferon γ and tumour necrosis factor α
    - cytokines activate dendritic cells in periphery
  - dendritic cells migrate to lymph nodes and inform helper T cells
    - that an infection is taking place
    - about the nature of pathogen
    - where is the pathogen invading
  - helper T cells orchestrate AI response

6.5 How does the adaptive immunity deal with virus:
- inside the cell
- outside the cell in the tissues or in the blood
- on mucosal surfaces, such as the lung or the intestine?

Viruses inside the cell

- Cytotoxic T cells (CTLs) kill cells which are infected with virus
  - CTLs can recognise the presence of virus derived molecules in infected cell
  - all cells display their contents as small peptides bound to major histocompatibility complex class I molecules on their surface
    - any cell displaying foreign peptides is killed by CTL
- Why do CTLs not recognise host cell peptides (i.e. self-peptides)?
  - any such CTLs are eliminated in the thymus during development

Viruses outside the cell

- Antibodies deal with extracellular virus
  - antibodies are proteins produced by plasma cells
    - plasma cells are derived from B cells
  - antibodies bind specifically and strongly to extracellular virus
- Three important classes of antibodies required to deal with extracellular virus
  - IgM
    - produced at the start of immune response
    - binds to virus and activates complement system
    - increases binding of complement C3b protein to viruses
    - stimulates lysis of enveloped viruses
    - prepares viruses for digestion by macrophages
  - IgG
    - produced later during the immune response by class switching
    - binds to virus and neutralises viral infectivity
    - activates complement system but not so efficiently as IgM
    - prepares viruses for digestion by macrophages (binds to the IgG receptor on macrophages and neutrophils)
    - IgG crosses placenta and protects new born
  - IgA
    - produced later during the immune response by class switching
    - secreted as a dimer onto mucosal surfaces

Viruses on mucosal surfaces (e.g. lung or intestine)

- IgA class of antibodies deals with virus on mucosal surfaces
  - secreted as a dimer, which allows IgA to:
    - survive proteases in intestinal tract and lungs
    - crosslink virus into clumps
    - neutralise viral infectivity
  - prepares virus to be eliminated in mucus
    - e.g. by coughing or cilary action
  - secreted in breast milk
    - protection of breast-fed children
- IgG class of antibodies can also play a role on mucosal surfaces

6.6 To which families of viruses do hepatitis B virus and hepatitis C virus belong? How do these viruses escape immune surveillance?

Hepatitis B virus (HBV)

- Member of hepadnavirus family
- Enveloped, icosahedral virus with a partially incomplete double stranded DNA genome

How does hepatitis B virus escape immune surveillance?

- Innate immunity
  - HBV released from infected cell by budding
    - infected cell not lysed, macrophages not aware of viral activity
  - only one strand of DNA genome codes for protein
    - low level of dsRNA, low level of interferon induction
  - HBV terminal protein interferes with interferon response
- Adaptive immunity
  - hepatitis B "virus-related particles" (VRPs) produced in large excess during viral infection
    - VRPs comprise viral glycoprotein HBsAg,
  - VRPs
    - are antigenically similar to virus particle
    - titrate out antibodies specific to HBV
  - antibodies against HBV only detectable after infection has been cleared
  - major defence against HBV during infection is therefore cytotoxic T cell response
    - acute infections cleared by this system
    - in chronic infections, cytotoxic T cell killing of infected hepatocytes is inefficient
    - due to reduction in synthesis of major histocompatibility complex class I molecules?
  - HBV does not escape adaptive immune system by mutation
    - primary structure of its surface glycoprotein remains constant
    - only one serotype of HBV found worldwide
- Stability of HBV surface glycoprotein important factor in success of HBV vaccine

Hepatitis C virus (HCV)

- Member of flavivirus family
- Enveloped, icosahedral virus with a single stranded RNA genome of positive sense

How does hepatitis C virus escape immune surveillance?

- Innate immunity
  - virus has envelope and replicates through double stranded RNA intermediate $\Rightarrow$ strong interferon inducer
  - but HCV encodes at least two proteins which disturb interferon response
- Adaptive immunity
  - HCV infection induces a strong B and T cell response
    - dramatically reduces the viral load in blood
  - error-prone viral RNA polymerase generates new genotypes which are not recognised by adaptive immunity
  - enables new wave of HCV synthesis which generates new adaptive immunity response
    - cycle of escape begins again
- This mutational strategy of hepatitis C virus makes prospect for a vaccine poor

6.7 Certain viruses interfere with the immune system of their host. Give two examples of such viruses and outline which components of the immune system are affected.

Two viruses which interfere with the host immune system

- Herpes simplex virus type 1
  - member of herpesvirus family
  - enveloped, icosahedral with a double stranded linear DNA genome
- Measles virus
  - member of paramyxovirus family
  - enveloped, helical virus with a single stranded RNA genome of negative sense

Herpes simplex virus type 1 (HSV-1) strategy

- HSV-1 reduces the synthesis of major histocompatibility complex (MHC) class I molecules in infected cells
  - MHC class I molecules present peptides from the cell to immune system (i.e. to cytotoxic T cells)
  - cytotoxic T cells sense presence of viral (foreign) peptides and lyse infected cells
  - reduction in MHC class I molecules reduces cytotoxic T cell killing
- Natural killer (NK) cells are very important in controlling all herpesvirus infections
  - NK cells survey cell surfaces
    - if both reduced numbers of MHC class I molecules and foreign molecules are detected as in HSV-1 infection, cell is killed
  - people with defects in NK cells especially at risk for infection with all herpesviruses
- HSV-1 also interferes with complement activity
  - protein on viral surface inactivates complement protein C3b
    - prevents lysis of virus by complement complex and phagocytosis by macrophages

Measles virus strategy

- Measles virus targets two key components of immune system
  - dendritic cells (DCs)
    - DCs under lung epithelia infected productively
    - DCs migrate to lymph nodes where virus is produced
  - macrophages
    - productively infected in lymph nodes
- Consequences of reduction of DCs and macrophages
  - dendritic cells (DCs)
    - no stimulation of adaptive immunity
  - macrophages
    - no destruction of extracellular virus
    - no stimulation of innate immunity or dendritic cells
  - immune response is lowered or incorrect
  - adaptive immunity may receive no or conflicting signals
- Medical importance
  - immunosuppression in measles infection
    - may last for a year
    - may allow secondary infections (e.g. pneumonia) which can be lethal
    - made worse by malnutrition

7 Prevention and control of viral diseases

7.1 Name five routes of transmission for viruses. For each route, give two examples of viruses transmitted in this way.

Respiratory route

- Inhalation of droplets and aerosols into the lungs
- Viruses transmitted include:
  - influenza A virus
  - measles virus

Oral-faecal route

- Certain viruses are excreted in faeces
  - viruses can then contaminate drinking water and be ingested to infect gastro-intestinal tract
  - poor personal hygiene can also contribute to ingestion of excreted viruses
- Viruses transmitted include
  - hepatitis A virus
  - human rotavirus
- Rotaviruses are the major causative agent of infant diarrhoea worldwide

Mother to child

- Viruses can be transmitted
  - across the placenta
  - during birth
  - by breast feeding
- Viruses transmitted include
  - rubella virus
    - across the placenta
  - human immunodeficiency virus type 1 (HIV-1)
    - during birth, breast feeding
- Rubella virus replicates in foetus and causes deformations

Close physical contact and sex

- E.g. kissing
  - Epstein-Barr virus
- E.g. sexual intercourse
  - HIV-1

Mechanical means

- Includes anything which causes penetration of the skin
- By arthropods (ticks, mosquitoes and sandflies)
  - yellow fever virus is transmitted by mosquito *A. aegypti*
- By needles and similar sharp objects for infusions, injections, transfusions and body decorations
  - e.g. hepatitis C virus

7.2 Illustrate with at least three examples how viral infections can be prevented.

Viral infections can be prevented by:

- Correct nutrition
  - strengthens immunity
- Correct hygiene
  - reduces viruses transmitted by oral-faecal route
    - e.g. hepatitis A virus
- Life style changes
  - reduce sexually transmitted viruses
    - e.g. herpes simplex virus type 2 (HSV-2)
    - e.g. human immunodeficiency virus type 1 (HIV-1)
- Vector control
  - reduces viruses transmitted by arthropod vectors
  - e.g. mosquito control reduces
    - dengue virus infections
    - yellow fever virus infections
- Vaccination
  - eradication of smallpox virus
  - control of many other viral diseases
- Anti-viral therapy in special cases
  - e.g. transmission of HIV-1 from mother to child

Three specific examples

- Dengue virus transmission
  - transmitted by mosquito *Aedes aegypti*
  - mosquitoes lay eggs preferentially in small pools of water, e.g.
    - in hollows of trees in Africa
    - in plastic containers and tyres in South America
  - improved disposal of plastic and tyres controls vector and leads to prevention of viral infection
- Measles virus
  - transmitted via respiratory route
  - viral infection leads to immunosuppression and occasionally (every 1:10,000 infections) to encephalitis
  - in Western countries, measles virus largely controlled by attenuated viral vaccine
  - vaccine not widely available in developing countries
    - still high rates of secondary infections and neurological problems
- Hepatitis B virus
  - hepatitis B virus endemic across Asia
    - virus transmitted within family in first years of life
    - either direct from mother to child or from siblings to child
    - infection in childhood leads to carrier state and high risk of progression to hepatocellular cancer (HCC)
  - prevention by subunit vaccine
    - hepatitis B surface glycoprotein produced in yeast
    - glycoprotein spontaneously assembles into virus-related particle
    - used as safe and effective vaccine
    - reduction in HCC rates in e.g. Taiwan evident

7.3 What are the advantages and disadvantages of live attenuated viral vaccines? How are such vaccines produced?

What are the advantages and disadvantages of live attenuated viral vaccines?

- Advantages
  - live attenuated viral vaccines mimic the natural infection without causing disease
    - virus can still replicate in host
    - inoculated virus is amplified
    - new virus particles are produced
  - generation of a B cell response and production of memory B cells
    - memory B cells protect from future infections
  - infection of cells leads to generation of cytotoxic T cell response
- Disadvantages
  - live attenuated viral vaccine may cause the very disease it should protect against because it may:
    - contain virulent virus (e.g. by reversion of attenuated strain)
    - recombine with related virus and generate virulent virus
    - still cause disease in immunosuppressed patients

How are live attenuated viral vaccines produced?

- Problem:
  - how to produce a live virus lacking virulence?
- Solutions:
  - adapt the virus to replicate in cells not derived from the natural host
  - adapt the virus to replicate in cells at lower temperature
  - in both cases, viruses may lose virulence but may still replicate in original host

Examples of attenuated viral vaccines

- Oral poliovirus vaccine
  - produced by passage of poliovirus intracerebrally in monkeys and in monkey cells
  - vaccine associated disease: 1 in a million doses
- Measles virus vaccine
  - produced by passage of measles virus in human and chicken cells
  - some passages at low temperatures
  - vaccine associated disease: encephalitis reported in fewer than 1 in a million vaccinated persons
- Examples of other live attenuated vaccines in use
  - mumps virus
  - rubella virus
  - yellow fever virus

7.4 Why are anti-bacterial compounds much more effective than anti-viral compounds? Give two examples of anti-viral compounds in use. Which problems can occur with the use of anti-viral compounds?

Why are anti-bacterial compounds more effective than anti-viral compounds?

- Both anti-bacterial and anti-viral compounds must interfere with a specific bacterial or viral target, but not with the host
- Bacteria are generally free-living organisms which possess their own systems for energy generation, protein synthesis, nucleic acid synthesis and cell wall construction
- These systems may not exist in the host or may be different from host systems
  - offer targets for attack by anti-bacterial substances
- In contrast, viruses:
  - replicate inside the host cell
  - use many systems from the host cell
  - encode only a small number of proteins
- Therefore,
  - targets for attack by anti-viral substances are limited
  - any potential compound must cross the host cell membrane

Two examples of anti-viral compounds in use

- Acyclovir
  - treatment of herpes simplex virus type 1 and 2, varicella-zoster virus
  - acyclovir is activated by viral thymidine kinase
  - activated compound has much higher affinity for viral DNA polymerase than for host DNA polymerase
  - compound preferentially incorporated into viral DNA where it prevents elongation of DNA chain
- Azidothymidine (AZT)
  - inhibits retroviral reverse transcriptase
  - used to treat human immunodeficiency virus type 1 (HIV-1)
    - above all used to treat HIV-1 seropositive women during and after pregnancy
    - prevents transmission from mother to child

Problems occurring in the use of anti-viral substances

- Emergence of viral mutants which are no longer susceptible to the substance
    - viral mutants arise because of large number of viral progeny and high mutation rates of certain viruses (e.g. HIV-1)
    - worsened by patients not taking substance regularly
        - permits some viral replication to occur which allows the selection of resistant mutants
- Anti-viral substances may not clear virus from the body
    - e.g. AZT can only work if reverse transcriptase is being made
    - proviral DNA which is not being replicated is not a target for AZT
    - anti-viral substances may need to be taken for life
- Many anti-virals have side-effects
    - made worse by need to be active in the cell
        - in contrast, anti-bacterials must be active inside the bacterium
    - made worse if anti-virals must be taken for life

7.5 Give four examples of viruses which can be transmitted from animals to humans. Name the natural host and the vector. How can infections with these viruses be prevented?

Examples of viruses which can be transmitted from animals to humans

- An infection transmitted from an animal to a human is termed a zoonosis
- Yellow fever virus
  - member of the flavivirus family
  - enveloped, icosahedral virus with a single stranded RNA genome of positive sense
  - natural host: primates and man
  - transmitting vector: mosquito (e.g. *Aedes aegypti*)
    - dengue virus is also transmitted by *A. aegypti*, but its natural host is only man
- Tick-borne encephalitis virus
  - member of the flavivirus family
  - enveloped, icosahedral virus with a single stranded RNA genome of positive sense
  - natural host: rodents
  - transmitting vector: ticks
- Rabies virus
  - member of the rhabdovirus family
  - enveloped, helical virus with a single stranded RNA genome of negative sense
  - natural hosts: foxes (Europe), coyotes and raccoons (North America), vampire bats (South America)
  - transmitting vector: none
    - transmission through bite from an infected animal
- Ebola virus
  - member of the filovirus family
  - enveloped, helical virus with a single stranded RNA genome of negative sense
  - causes haemorrhagic fever in humans
  - natural host: not clear, perhaps primates
  - transmitting vector: not identified
    - no arthropod vector
    - possible transmission through contact with an infected animal

Prevention of infection

- Vaccination
  - humans can be vaccinated against:
    - yellow fever virus (live attenuated vaccine)
    - tick-borne encephalitis virus (killed vaccine)
    - rabies virus (killed vaccine), can be given after bite by suspect animal
  - animals can be vaccinated against:
    - rabies virus (live vaccinia virus expressing rabies virus glycoprotein)
    - given as bait for wild foxes in Europe and coyotes in North America
- Vector control
  - yellow fever virus
    - removal of habitat for mosquitoes
    - pesticides during epidemics
- At present, no measures to prevent Ebola virus infections
  - but geographical range of its unknown host appears to limit spread

7.6 Name two:
- live viral vaccines
- killed whole viral vaccines
- viral vaccines containing only viral subunits or a viral protein

Why are live vaccines preferred? What disadvantages can live vaccines have?

Live viral vaccines

- Oral poliovirus vaccine
    - live attenuated poliovirus
    - one strain for each of the three poliovirus serotypes
- Measles virus vaccine
    - live attenuated measles virus
    - one strain for the one measles virus serotype

Killed viral vaccines

- Hepatitis A virus vaccine
    - inactivated hepatitis A virus
    - one strain for the one hepatitis A serotype
- Rabies virus vaccine
    - inactivated rabies virus
    - one strain for the one rabies virus serotype

Viral vaccines containing only viral subunits or a viral protein

- Influenza virus vaccine
    - influenza virus treated with detergent
    - releases haemagglutinin (HA) and neuraminidase (NA) proteins
    - vaccine contains HA and NA proteins from:
        - two influenza A virus strains
        - one influenza B virus strain
    - composition of vaccine adjusted each year
- Hepatitis B virus vaccine
    - hepatitis B surface antigen produced in yeast
    - surface antigen (glycoprotein) self-assembles to virus-related particle
    - one protein for the one hepatitis B virus serotype

Advantages of live viral vaccines

- Amplify the inoculated virus
- Mimic a real viral infection
- Produce both B cell and T cell response
- Produce long-lasting immunity

Disadvantages of live viral vaccines

- Attenuated viruses may revert to virulent strain
- The vaccine may be contaminated with virulent virus
- The vaccine strain may recombine with other viruses to generate virulent virus
- The vaccine may spread and infect non-vaccinated individuals
    - may cause disease, e.g. in immunosuppressed individuals

7.7 Which treatments are available for a human immunodeficiency virus type 1 (HIV-1) infection? How can infection with HIV-1 be prevented? Why is it proving so difficult to develop an effective vaccine against HIV-1?

Human immunodeficiency virus type 1 (HIV-1)

- Member of retrovirus family
- Enveloped, icosahedral virus with a single stranded RNA genome of positive sense
- RNA genome replicated through a DNA intermediate (proviral DNA)

Treatment of HIV-1 infections

- Using drugs which inhibit HIV-1 encoded enzymes, such as:
  - reverse transcriptase inhibitors
    - nucleoside analogues (e.g. azidothymidine (AZT))
    - non-nucleoside analogues (e.g. nevirapine)
    - strategy: to prevent conversion of genomic RNA into proviral DNA
  - protease inhibitors
    - e.g. saquinavir
    - e.g. indinavir
    - strategy: to prevent maturation of new virus particles into fully infectious ones
- Anti-HIV-1 compounds often taken together
  - termed "triple therapy" or "HAART" (highly active anti-retroviral therapy)
  - treatment must be taken for life
    - does not cure HIV-1 infection
- Recently, compounds developed which prevent HIV-1 entry into cell
  - efficacy still being tested

HIV-1 infections can be prevented by

- Screening of blood donors
  - exclude donors practising unsafe sex
- Inactivation of HIV-1 in blood and blood products by heat
- Preventing re-use of hypodermic needles
- Changing sexual behaviour
  - using condoms, reducing number of partners, etc.
- Prophylactic treatment with AZT to prevent mother to child transmission
  - before and during pregnancy
  - using alternatives to breast-feeding

A vaccine against HIV-1 is difficult to develop because

- Unclear which type(s) of immune response will be necessary to protect against HIV-1
  - during initial HIV-1 infection, strong cytotoxic T cell response and rapid production of HIV-1 specific antibodies
  - nevertheless, not sufficient to clear the virus from the body
- Existence of several subtypes of HIV-1 coupled with a high rate of mutation
- Unclear which HIV-1 proteins could provide a protective response
  - e.g. surface envelope glycoprotein gp120 is heavily gycosylated
  - conserved epitopes are hidden
    - these epitopes only shown during conformational changes occurring on entry
- Lack of suitable animal models
  - ethics of vaccine testing in humans
- Inacceptability of attenuated HIV-1 vaccines
  - may revert to wild-type virus
- Vaccine could generate inappropriate response to HIV-1
  - enhance viral infection instead of preventing it

7.8 How can flaviviruses be transmitted? Give examples. Why is there no effective vaccine against dengue virus? How can this virus still be controlled?

Flaviviruses

- Enveloped, icosahedral viruses with a single stranded RNA genome of positive sense
- Family includes several important human and animal pathogens

How can flaviviruses be transmitted?

VIRUS	TRANSMISSION
yellow fever virus dengue virus West Nile virus Japanese encephalitis virus	mosquitoes (arthropod borne viruses = arboviruses)
bovine viral diarrhoea virus hog cholera virus	oral-faecal
hepatitis C virus	blood transfusion drug abuse origin of ca. 50% infections unclear very low levels of sexual and mother to child transmission

Why is there no effective vaccine against dengue virus?

- Dengue virus has four serotypes
  - person suffers more severe disease if infected subsequently with a second serotype
  - inactivated viral vaccines for each serotype have been developed
    - protect against only one serotype
  - vaccine may not only fail to protect against another serotype, it may actually worsen the disease (potentiation)
- The mechanism for potentiation is unclear
  - antibodies may enhance entry of dengue virus
- Quadrivalent vaccine against all four serotypes is in clinical trials

Control of dengue virus by vector control

- Dengue virus transmitted by *Aedes aegypti* mosquito
- Mosquitoes preferentially lay eggs in small pools of water
  - e.g. abandoned plastic containers, tyres, etc.
- Control by:
  - removal and correct disposal of such articles
  - drainage of standing water near human habitation
- *A. aegypti* is a day-time feeder
  - control with impregnated mosquito nets not possible

7.9 Which possibilities are available to prevent and control viral infections? Describe the development and properties of the attenuated poliovirus vaccine.

Prevention and control of viral infections

- Proper nutrition
- Hygiene and sanitation
- Vector control
- Life-style and behavioural changes
- Vaccination
- Anti-viral drugs

Poliovirus

- Member of picornavirus family
- Non-enveloped, icosahedral virus with a single stranded RNA genome of positive sense

Development of attenuated oral poliovirus vaccine (OPV)

- OPV contains attenuated (non-virulent) strains of all 3 serotypes of poliovirus
  - attenuated strain for type 2 from naturally occurring isolate
  - attenuated strains for types 1 and 3 obtained by serial passage in different cells, tissues and organisms
  - e.g. serial passages for type 3 poliovirus
    - intracerebrally in monkeys
    - in vitro in monkey testicle cells
    - in vitro in monkey kidney cells
- Attenuation leads to loss of neurovirulence in humans
- However, virus can still replicate in intestine and generate immunity

Properties of OPV

- Trivalent
  - contains all 3 serotypes
- Attenuated
  - virus replicates in intestine, but does not replicate in neurons
  - no paralysis, no poliomyelitis
- Immune response
  - live vaccine, advantages therefore include
    - amplification of input virus
    - infection of cells and virus particle production
    - production of both B and T cell response
    - production of memory B cells
- Disadvantages
  - temperature sensitive, must be kept cold
  - reversion to neurovirulent phenotype
  - reversion can occur within 48 hours in infant gut
    - in type 3, one mutation mainly responsible for attenuated phenotype
  - vaccine induced disease
    - either from vaccine itself (1 case of poliomyelitis for 5×10^6 doses)
    - or from recombination in the wild with related enteroviruses

7.10 Why was it possible to eradicate smallpox virus? Why is it proving more difficult to eradicate poliovirus and measles virus?

Smallpox virus

- Member of poxvirus family
- Enveloped virus with complex particle and double stranded linear DNA genome

Why was it possible to eradicate smallpox virus?

- Virus properties
  - only one, antigenically stable serotype of smallpox virus
  - humans are only host for smallpox virus
  - no persistent or subclinical infections
  - infected person transmits smallpox virus only after appearance of symptoms
  - possible to interrupt chain of transmission by isolating infected persons and their contacts
- Vaccine properties
  - vaccine (vaccinia virus) stable
    - no reversion
    - no recombination
  - single vaccination left scar
    - i.e. permanent record
- Worldwide recognition that
  - smallpox was a life-threatening disease
  - smallpox caused large economic losses
  - financial investment in vaccination worthwhile

Why is it proving more difficult to eradicate poliovirus and measles virus?

- Similar to smallpox virus,
  - man is the only natural host for both viruses
  - measles virus has one antigenically stable serotype
  - poliovirus has just three antigenically stable serotypes
- In contrast to smallpox virus
  - both viruses can be spread by infected persons before symptoms occur
    - person may also be infected with poliovirus but not show symptoms (i.e. subclinical infection)
  - persistent infections
    - immunosuppressed persons may excrete poliovirus over long time periods
    - persistent infections with measles virus rare but possible
  - vaccines
    - poliovirus and measles virus vaccines must be kept cold
  - poliovirus vaccine
    - several doses of live poliovirus vaccine required
    - vaccinated persons excrete live poliovirus
    - attenuated live poliovirus vaccine may recombine with related viruses to become virulent
  - measles virus vaccine
    - measles virus vaccine may not protect completely
    - measles virus vaccine cannot be taken orally
  - no world-wide recognition that measles virus is life-threatening

7.11 How do vaccines induce protection against viral infections? Explain the different properties of IgM, IgG and IgA antibodies.

How do vaccines induce protection against viral infection?

- Vaccines mimic a viral infection without causing disease
  - adaptive immunity sets up an immune response
  - some B and T cells develop to memory cells
    - "remember" encounter and protect against infection with pathogenic virus
- Three main types of viral vaccines
  - live viral vaccines
  - inactivated viral vaccines
  - subunit viral vaccines
- All three types generate memory B cells
  - all viral vaccines elicit antibody production
  - live or inactivated virus particles and subunits are ingested and degraded by macrophages and dendritic cells
    - digestion products are presented to T cells
  - helper T cells activate B cells specific for invader
    - most activated B cells become plasma cells and produce antibodies
    - some develop to memory cells
- Memory B cells are reactivated when pathogenic virus invades
  - reactivation requires helper T cells
  - reactivation is rapid and occurs at low concentration of invader
- Memory B cell acts itself as antigen presenting cell
  - no need for macrophages or dendritic cells
  - faster response
- Only live viral vaccines generate memory T cells
  - cytotoxic T cell response only occurs when the cells are infected
  - inactivated and subunit viral vaccines cannot infect cells
- Memory B cells are sufficient for protection
  - indicated by success of inactivated and subunit viral vaccines
    - e.g. inactivated vaccines against hepatitis A virus, poliovirus
    - e.g. subunit vaccines against hepatitis B virus, influenza virus

Properties of IgM, IgG and IgA antibodies

- IgM antibodies
  - first class of antibody synthesised
  - systemic
  - oligomeric structure
    - high potential for cross-linking antigens
  - when IgM is bound to antigen, activates complement
    - IgM-antigen complex enhances binding of complement proteins to viral surfaces
- IgG antibodies
  - produced by class switching
  - systemic
  - bind to antigen and prepare it for digestion by macrophages and neutrophils
  - poor activators of complement
- IgA antibodies
  - produced by class switching
  - secretory
    - protect mucosal surfaces e.g. intestine, lung, nasal cavity
  - do not activate complement
  - do not prepare antigens for digestion
  - synthesised as covalent dimer
    - promotes crosslinking of antigens
    - leads to formation of clumps which can be washed away (e.g. by ciliary action in lungs)

7.12 Which human pathogens are found in the herpesvirus family? How does the anti-viral compound acyclovir work?

Human pathogens in the herpesvirus family

- Enveloped, icosahedral viruses with a linear double stranded DNA genome

VIRUS	DISEASE
herpes simplex virus 1 (HSV-1)	• labial herpes (cold sores)
herpes simplex virus 2 (HSV-2)	• genital herpes (genital sores and ulcers)
varicella-zoster virus (VZV)	• chickenpox • shingles (adults) • reactivated in immunosuppression
Epstein-Barr virus (EBV)	• subclinical in infants • infectious mononucleosis in teenagers ("kissing disease") • factor in Burkitt's lymphoma
cytomegalovirus (CMV)	• subclinical if contracted in infancy or adulthood • contraction during pregnancy may cause foetal abnormalities • risk in immunosuppression (e.g. bone marrow transplantation, HIV-1 infection)
human herpesvirus 6 (HHV-6)	• rash in children
human herpesvirus 8 (HHV-8)	• factor in Kaposi's sarcoma • specially common in people suffering from AIDS

- Most important clinical characteristic of herpesvirus infections is latency
  - life-long infections which can reoccur at any time

How does the anti-viral compound acyclovir work?

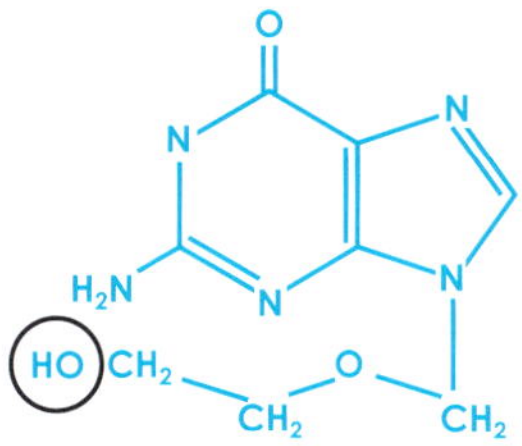

- Acyclovir is a nucleoside analogue and actually a prodrug
- Must be converted to nucleotide triphosphate using ringed hydroxy group
  - equivalent to 5′ hydroxy group in a nucleoside
- Can then be incorporated into a DNA chain
- However, no equivalent of 3′ hydroxy group for extension
  - growing DNA chain terminated
- Used to treat HSV-1, HSV-2 and VZV infections

Conversion to triphosphate

- Occurs in two steps
  - conversion to monophosphate by herpesvirus thymidine kinase
    - specific for herpesvirus infected cells
  - conversion to triphosphate by cellular enzyme

Incorporation into DNA chains

- By both viral and cellular DNA polymerases
  - however, specificity for viral enzyme 10 fold higher
  - second level of specificity for virus infected cells

Points to note

- Only affects cells in which virus is actively replicating
- In use for 20 years, but nevertheless, low frequency of:
  - side-effects
  - HSV strains resistant to acyclovir

7.13 Give an example of a:
- live viral vaccine
- killed viral vaccine
- subunit viral vaccine

Which type of immune response do they generate? What are the advantages and disadvantages of each type of vaccine?

Vaccine examples

- Live viral vaccine - vaccinia virus (against smallpox virus)
- Killed viral vaccine - inactivated poliovirus (the Salk vaccine)
- Subunit viral vaccine - human papillomavirus
  - licensed since 2006
  - contains only capsid protein VP1
  - VP1 self assembles to highly immunogenic virus-like particles

Which type of immunity do different vaccines generate?

- Live viral vaccines generate both a B cell and a T cell response
- Killed and subunit vaccines generate only a B cell response
- All types of viral vaccines induce memory B cells
  - mainly responsible for protective effect of vaccine

What are the advantages and disadvantages of each vaccine type?

- Live viral vaccines
  - advantages
    - mimic viral infection, produce B and T cell response
    - inoculated virus replicates and is thus amplified
    - long lasting immunity possible with one dose
  - disadvantages
    - may spread and cause disease in immunosuppressed patients
    - attenuated viruses may regain virulence by reversion or recombination
    - vaccine may be contaminated with wild-type virus
    - vaccine may need to be kept cool to remain effective
- Killed viral vaccines
  - advantages
    - cannot spread to immunosuppressed patients
    - cannot revert by recombination or reversion
    - vaccine not so sensitive to temperature
  - disadvantages
    - no T cell response
    - no amplification of inoculated virus, therefore one injection does not sufficiently stimulate the immune system
    - several injections may be required
    - maybe contaminated with wild-type virus if inactivation is incomplete
    - if inactivation is too harsh, it may destroy the virus, preventing a protective response
- Subunit vaccines
  - advantages
    - cannot cause disease
    - must not be kept cool
  - disadvantages
    - no T cell response
    - no amplification, therefore one injection does not sufficiently stimulate the immune system
    - several injections may be required

8 Evolution, emerging viruses and other agents

8.1 State three mechanisms through which viruses can evolve. Provide an example for each mechanism.

Three mechanisms of viral evolution

- Mutation – e.g. human immunodeficiency virus type 1 (HIV-1)
- Recombination – e.g. poliovirus
- Reassortment – e.g. influenza virus

Mutation

- Virally encoded RNA dependent RNA polymerases have no proofreading and are therefore likely to make errors
- Retrovirus encoded reverse transcriptase (RT: RNA and DNA dependent DNA polymerase) also has no proofreading
  - HIV-1 RT makes on average one error per every 10,000 bases synthesised
  - i.e. one mistake per genome (10^4 nucleotides) copied
    - i.e. each time a HIV-1 particle infects a cell
  - in an AIDS patient, 10^9 particles synthesised per day
    - theoretically mutations in every base can be generated
- Most DNA viruses use cellular DNA dependent DNA polymerases
  - nevertheless, higher mutation rates found in viral DNA than in cellular DNA
- Production of large numbers of progeny ensues production of viable virus despite high error rates
- Medical importance:
  - generation of HIV-1 strains resistant to anti-viral therapy

Recombination

- Occurs in both RNA and DNA viruses when two different strains infect same cell
    - may also occur in retroviruses between two genomic RNAs of one virus
- Allows viable virus to be generated from two defective genomes
- E.g. poliovirus (PV)
    - PV strain defective at 5' end (X)

    - PV strain defective at 3' end (Y)

    - viable PV recombinant - both defects removed

- Medical importance
    - recombination of PV vaccine strains with wild-type PV or related enteroviruses in the environment
        - e.g. outbreak of poliomyelitis on Hispaniola 2000/01

Reassortment

- Occurs in influenza virus when two different strains infect the same cell
- Hybrid virus generated which contains RNA gene segments from both viruses
    - referred to as reassortant
- Medical importance
    - antigenic shift in human influenza virus
        - human influenza virus may acquire RNA segment coding for haemagglutinin from influenza virus circulating in birds
        - reassortant virus is new to human immune system and may cause epidemic of influenza

8.2 How can viruses evolve? Give two examples of viruses which recently emerged through viral evolution.

How can viruses evolve?

- Two fundamental characteristics allow viruses to evolve:
  - viruses have high mutation rates during replication
    - polymerases from RNA viruses have no proofreading activity
    - genomes of DNA viruses show higher rates of mutation than cellular DNA even when the same cellular DNA polymerase is used
  - viruses produce large numbers of new viruses (progeny)
    - e.g. hepatitis B virus infection may lead to production of 10^{11} new virus particles per day
- In addition, novel combinations of viral genomes arise by:
  - recombination
    - may occur when two different strains of a virus infect the same cell
    - may occur during reverse transcription of two genomic RNA molecules of a retrovirus
    - third strain may arise whose genome comprises parts of other two
  - medical importance:
    - e.g. vaccine strains of poliovirus may recombine with other human enteroviruses
    - e.g. recombination between subtypes of HIV-1
  - reassortment
    - only in viruses with segmented genome
    - occurs when two different strains infect the same cell
    - third strain may arise which contains segments from the other two
  - medical importance:
    - e.g. strains of influenza virus with a new antigenicity may be produced
    - e.g. recombination amongst rotavirus strains (childhood diarrhoea)

Two viruses which recently emerged through viral evolution

- Canine parvovirus
  - causes inflammation and loss of intestinal wall in dogs
  - predecessor was feline parvovirus which causes panleukopenia in cats
  - two mutations in feline parvovirus sufficient to allow the virus to infect dogs
  - further evolution of canine parvovirus made it a distinct virus which is no longer able to infect cats
- Influenza virus
  - constant mutation (antigenic drift) and reassortment (antigenic shift) give rise to influenza outbreaks in humans, birds and pigs across the world
    - e.g. human influenza pandemics of 1957 and 1968
    - e.g. outbreak of swine influenza in USA in 1998 to 2000
    - e.g. current pandemic of avian influenza

8.3 To which family does the virus which causes SARS belong? Sketch the virus particle and the genome of this family and indicate the individual components. Outline the mechanisms used to express the genetic information.

Coronaviruses

- Virus which causes SARS (severe acute respiratory syndrome) belongs to the coronavirus family
- Enveloped, helical viruses with a single stranded RNA genome of positive sense

The coronavirus particle

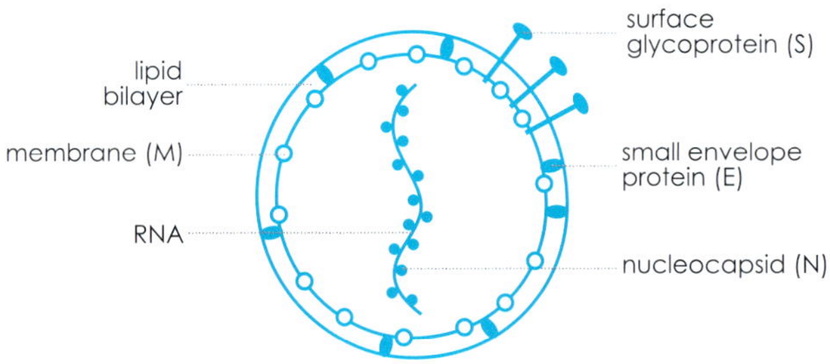

- Surface glycoprotein (S) is responsible for attachment to host cells and for inducing neutralising antibodies
- RNA genome comprises about 30 000 nucleotides
  - largest viral RNA genome known
  - helical capsid allows large RNA genome

Coronavirus genome

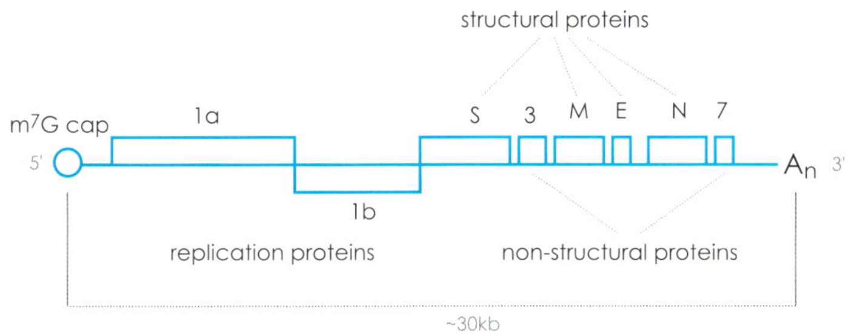

Mechanisms used to express the genetic information

- RNA is of positive sense and can be translated immediately on release of viral RNA into cytoplasm
  - open reading frames (ORFs) 1a and 1b are translated from same start codon (AUG)
  - translating ribosomes can either:
    - terminate protein synthesis at stop codon of ORF 1a
    - or shift into another ORF to synthesise extended protein 1a-1b
    - mature proteins generated by proteolytic processing of 1a and 1a-1b
- Product of open reading frame 1b is viral polymerase
  - viral polymerase can synthesise:
    - full-length negative copies of viral genome
    - sub-genomic negative RNAs corresponding to ORFs for S, E, M, N and non-structural proteins
  - full-length negative copies are used as a template for synthesis of full-length positive RNA
    - can be packaged into new virus particle
  - sub-genomic negative RNAs are used to synthesise sub-genomic positive sense mRNAs which are translated to generate the encoded proteins
- Coronaviral strategy for genetic expression:
  - direct translation of viral RNA
  - ribosomal frame shifting
  - proteolytic processing
  - synthesis of sub-genomic mRNAs

8.4 How can a new virus "emerge"? Use poliovirus to support your answer.

How can a new virus "emerge"?

- Emerging viruses must evolve from existing viruses
  - viruses must have ancestors
- Three main mechanisms allow viruses to emerge:
  - existing virus evolves to infect an essentially immune population
    - natural host has no immunity to new variant
    - e.g. influenza virus
  - existing virus enters a naïve population
    - e.g. West Nile virus emerging in USA affecting human, horse and bird populations
  - existing virus evolves to enter new host species
    - e.g. ancestor of canine parvovirus was feline parvovirus
    - two mutations in feline parvovirus
    - mutated virus capable of infecting dogs, but no longer able to infect cats

Emergence of poliovirus and poliomyelitis in developed countries

- Poliovirus
  - member of picornavirus family
  - non-enveloped, icosahedral virus with a single stranded RNA genome of positive sense
  - transmitted by oral-faecal route
- Before ca. 1900
  - poliovirus present in water supplies and sewage
    - some protection from maternal antibodies
    - but most people infected in early infancy
    - at this age, poliovirus infection is subclinical
  - infection in infancy gives life-long protection
- After ca. 1900
  - improvements in hygiene, better water supplies and sewage system
    - no exposure to poliovirus in infancy
    - no "natural vaccination"
  - exposure to virus now in adolescence or adulthood
    - poliovirus infections again mostly subclinical
    - but much higher frequency of paralysis (i.e. poliomyelitis)
  - virus emerged by entering naïve population
  - naïve population created by modern sanitation

8.5 How do humans provide new ecological niches for viruses? Why can viruses evolve so rapidly to use the opportunities presented? Which parameters limit their evolution?

Humans provide new ecological niches for viruses by:

- Movement
  - movement into new areas previously not visited by humans
    - removal of forest in Africa appears to have allowed Ebola virus access to humans
  - world-wide travel
    - international shipping spread influenza virus around the world in 1918/19
    - international air travel spread SARS virus in 2003
- Contact with animals
  - proximity of poultry, pigs and humans
    - spread of previously avian-specific strains of influenza virus to humans in Eastern Asia in 1997 and 2003-2005
    - mixture of many animals in Chinese markets offered SARS virus close proximity to humans
- Behavioural changes
  - increase in oral sex practices
    - "genital" herpes simplex virus type 2 now found more often in oral body parts
  - desire to eat fresh exotic fruit
    - contaminated produce from Mexico caused outbreak of hepatitis A virus in Eastern USA
- Providing of new opportunities for viral vectors
  - flooding of valleys by dams
    - mosquitoes can expand into flooded areas
    - spread of Rift Valley virus
  - failure to dispose of plastic canisters and tyres
    - provide new habitats for mosquito *A. aegypti*
    - leads to outbreaks of dengue virus in South America (e.g. Brasil)

Evolution of viruses is driven by

- Mutation rate
  - viruses have high rate of mutation
  - RNA viruses make one mutation per every 10,000 nucleotides synthesised
  - DNA viruses are more accurate but still mutate at higher rate than host
- Enormous number of progeny
  - e.g. AIDS patient synthesises 10^9 virus particles of HIV-1 per day
  - e.g. hepatitis B virus patient produces 10^{11} virus particles per day

Evolution of viruses is limited by

- Size of icosahedral capsid
  - only a certain amount of genetic material can be accepted
- Only a limited number of ways to decode genetic information
- Mutations may adversely affect activity of viral proteins in the cell
- Too many mutations will inactivate the virus
- An extremely virulent virus will kill its host and thus itself

8.6 How can viruses emerge without genetic change?

How can viruses emerge without genetic change?

- Viruses which emerge without genetic change are generally taking advantage of a new niche presented to them by human activity
  - numerous examples, including:
- Enabling humans or farm animals to enter previously inaccessible areas
  - e.g. clearing of rain forest or jungle
  - allowed human contact with gorillas in Africa
    - outbreaks of Ebola virus in humans
  - allowed contact of pigs with bat urine in Malaysia
    - spread of Nipah virus to pigs
    - subsequent spread of Nipah virus to humans
- Movement of infected human, animals or viral arthropod vectors across large distances
  - allows viruses to access naïve populations
  - movement of humans during colonisation of Americas
    - spread of smallpox and measles viruses among native American Indians
  - movement of humans in 1918/19
    - spread of influenza virus from USA to Europe and then to the rest of planet
  - movement of birds or mosquitoes carrying West Nile virus from Africa to North America at the end of 20th century
    - emergence of West Nile virus in birds, horses and humans
  - movement of animals from Africa to USA recently
    - bought as exotic pets
    - spread of monkeypox virus from Africa to USA
  - an example not caused by human intervention
    - outbreak of Usutu virus in Vienna (2001) amongst blackbirds
    - caused by migration of infected birds from South Africa to Vienna
    - virus transmitted by mosquitoes in Vienna
    - Usutu virus does not cause disease in humans

- Human behavioural and life-style changes
    - introduction of water supply and sewage programmes in USA and Europe
        - prevented exposure and thus immunisation of young children to poliovirus
        - instead, exposure to poliovirus occurred as young adult or adolescent (e.g. in swimming pools)
        - poliovirus more likely to cause poliomyelitis at this age
        - human activity broke "natural" chain of immunisation
    - provision of opportunities for vector expansion
        - building dams or incorrect disposal of tyres and plastic containers
        - provide new or preferable breeding grounds for mosquitoes carrying Rift Valley virus or dengue virus
    - changes in human sexual behaviour

8.7 Explain what is meant by antigenic shift and antigenic drift in influenza virus.

Influenza virus

- Member of the orthomyxovirus family
- Enveloped, helical virus with a segmented single stranded RNA genome of negative sense
- Influenza virus has two ways of escaping immune surveillance:
    - antigenic shift
    - antigenic drift

Antigenic shift

- Viral proteins haemagglutinin (HA or H) and neuraminidase (NA or N) determine antigenicity of influenza virus
- Influenza virus strain designated by HA and NA subtypes
    - for influenza A virus, 15 HA and 9 NA subtypes are known
    - HA subtypes 1, 2 and 3 can cause epidemics in humans
        - e.g. 1918/19 pandemic strain was H1N1
    - individual who has recovered from or who is vaccinated against H1N1 is protected against that strain, but not against H2N1 or H3N3 etc.
- Subtype changes can occur when two different influenza viruses infect same cell
    - e.g. H1N1 strain growing well in humans and H3N2 strain growing well in birds infect pig at the same time
        - gene segments can be reassorted so that a new combination such as H3N1 can arise
    - human population may have no immunity against this subtype
        - a shift in the antigenicity has occurred

Antigenic drift

- Within a subtype, antigenicity changes over time
  - caused by introduction of point mutations during replication by error-prone viral RNA polymerase
    - viral RNA polymerase lacks proofreading activity
  - HA can tolerate these changes but still retain the ability to recognise sialic acid
    - influenza virus can still attach to host cell
  - mutations may mean that the immune system no longer recognises strain although the subtype has not changed
    - referred to as antigenic drift
  - to account for antigenic drift, influenza virus vaccines are adjusted yearly
    - calculated to provide immunity against currently circulating strains

Medical importance

- Concern that HA subtypes currently circulating in birds (e.g. H5, H7) may spread from human to human
  - human population would be immunologically naïve
- But: all human pandemics of 20th century were H1, H2 or H3 subtypes

8.8 What are transmissible spongiform encephalopathies (TSEs)? How do these diseases arise and what are thought to be causative agents? How are these causative agents thought to replicate?

What are transmissible spongiform encephalopathies (TSEs)?

- TSEs are a group of neurodegenerative diseases with a long incubation time
  - lead to behavioural and neurological changes including dementia and eventually death
  - on autopsy, brain has a characteristic sponge-like appearance
- TSEs include:
  - scrapie (sheep)
  - bovine spongiform encephalopathy (BSE, cattle)
  - Creutzfeld-Jakob disease (CJD, humans)
  - new variant Creutzfeld-Jakob disease (vCJD, humans)
  - Kuru (humans)

How do TSEs arise?

- Generally a genetic disorder
  - predisposition caused by presence of certain genotypes in germ-line
  - somatic mutation during life may lead to a change in a certain protein
- Transmission from affected individuals
  - using human growth hormone preparations made from human hypophysis obtained from individuals with CJD
  - eating infected tissues
    - e.g. sheep eating grass contaminated with remains of placenta from a sheep with scrapie
    - e.g. cattle eating bone meal containing material from cattle with BSE
    - e.g. humans eating meat from cattle with BSE led to appearance of new-variant Creutzfeld-Jakob disease (vCJD)

What are thought to be the causative agents of TSEs?

- Prions are thought to be causative agents of TSEs
  - strong body of evidence that they are infectious proteins
  - but absence of nucleic acid not conclusively shown
- Prions can exist in two forms
  - normal, non-pathogenic, protease-sensitive form – termed PrP^{c} (c = cellular)
  - abnormal, pathogenic, protease-resistant form – termed PrP^{sc} (protease resistant protein found in scrapie)
- PrP^{c} and PrP^{sc} are different conformations of the same protein
  - conformational change from PrP^{c} to PrP^{sc} occurs very slowly and is irreversible
    - PrP^{sc} catalyses the conversion of PrP^{c} to PrP^{sc}
  - mutations in PrP^{c} accelerate formation of PrP^{sc}
    - accelerates development of TSEs
- Protease-resistance of PrP^{sc} explains its transmissible nature
  - allows PrP^{sc} to survive the digestive tract
  - PrP^{sc} may pass to brain where it can catalyse conversion of PrP^{c} in brain of new individual

Medical importance

- Transmission via growth hormone eliminated by use of recombinant growth hormone
- BSE being eliminated by stopping the feeding of processed animals to cattle
- Scrapie reduced by out-breeding to avoid certain PrP genotypes
- Genetic source of TSEs will always be present at a low incidence
  - TSEs can always emerge in populations of supposedly disease-free animals

9 Emerging Questions

9.1 Which viruses package their viral polymerase into the viral particle? What is the significance of their presence in the particle? How can viruses that do not package their polymerase in the particle copy their genetic material?

Viruses that package their viral polymerase into the particle

- Negative strand RNA viruses
  - viral RNA polymerase required immediately after release of viral genome to synthesise positive sense sub-genomic RNAs
  - later, polymerase synthesises full-length negative strand RNAs
    - measles virus (paramyxovirus)
    - Ebola virus (filovirus)
    - rabies virus (rhabdovirus)
    - influenza A virus (orthomyxovirus)
- Double stranded RNA viruses
  - viral RNA polymerase synthesises full-length positive sense RNAs in the virus particle
  - RNA extruded from particle into cell during synthesis
  - double stranded RNA never found in infected cell
    - rotavirus (reovirus)
    - blue tongue virus (reovirus)
- Retroviruses
  - reverse transcriptase copies viral RNA into proviral DNA in the cytoplasm
    - HIV
    - avian leukosis virus
- Hepadnavirus
  - polymerase copies viral RNA into double strand DNA in the assembling viral particle
  - polymerase remains covalently bound to minus strand DNA
    - hepatitis B virus
- Poxviruses
  - viral DNA polymerase required for viral DNA synthesis in cytoplasm
  - poxviruses are the only DNA viruses that encode all enzymes for DNA replication
    - smallpox virus
    - monkeypox virus

Viruses that do not package their viral polymerase into the particle

- Positive strand RNA viruses
  - viral genome translated immediately on release to generate viral proteins, including viral RNA polymerase
    - hepatitis C virus (flavivirus)
    - foot-and-mouth disease virus (picornavirus)
    - sindbis virus (togavirus)
    - SARS virus (coronavirus)
- Most DNA viruses
  - either use host DNA polymerases
    - human papillomavirus
    - simian virus 40 (polyomavirus)
  - or viral DNA polymerase encoded on viral genome and synthesised during replication
    - adenoviruses
    - herpesviruses

Concept

- all virally encoded polymerases, whether they are packaged or not perform reactions that cellular polymerases cannot

9.2 How do influenza A viruses bind to cells? Why do certain influenza A viruses bind specifically to avian cells and others to human cells? How can the highly pathogenic H5N1 avian strain infect humans? How did the 2009 pandemic strain of influenza A virus arise?

How do influenza A viruses bind to cells?

- Influenza A virus
  - member of the orthomyxovirus family
- Haemagglutinin (HA) protein in viral envelope responsible for viral attachment to cells
  - binds specifically to a sugar molecule on the surface of the cell called sialic acid (also called N-acetylneuraminic acid)
  - sialic acid is in turn covalently linked to a chain of sugar residues connected to a protein anchored in the cell membrane
    - sialic acid is always the outermost sugar on the chain
- Binding of influenza A viruses to avian and/or human cells depends on linkage of sialic acid to the next sugar in the chain

Why do certain influenza A viruses bind specifically to avian cells and others to human cells?

- In epithelial cells of the avian intestinal tract, sialic acid is usually linked to galactose in an α-2, 3 glycosidic linkage
  - avian specific influenza A viruses recognise predominantly this structure
    - e.g. highly pathogenic H5N1 strain and currently circulating avian specific H1N1 strains
- In cells of human upper respiratory tract, sialic acid is usually connected to galactose in an α-2, 6 glycosidic linkage
  - human specific influenza A viruses recognise predominantly this structure
    - e.g. currently circulating human specific H1N1 strains, human specific H3N2 strains
  - some influenza A virus strains bind to both α-2, 3 and α-2, 6 glycosidic linkages
    - e.g. 1918 H1N1 pandemic strain

How can avian highly pathogenic H5N1 influenza A virus infect humans?

- Cells of human lower respiratory tract do possess α-2, 3 glycosidic linked sialic acid
  - for infection to occur, large amount of virus has to reach lower respiratory tract
    - only possible on close contact with infected birds or siblings
  - virus release from infected person not efficient, coughing not sufficient to expel virus from lung
    - low amount of released virus prevents human to human transmission

How did the 2009 pandemic strain of influenza A virus arise?

- 2009 influenza A virus pandemic strain designated swine-origin H1N1 influenza virus
  - arose by reassortment of gene segments from four different viruses
    - classical swine influenza A virus
    - North American avian influenza A virus
    - human influenza A virus
    - European avian-like swine influenza A virus
  - reassortment of first 3 viruses occurred before 1997
    - probably took place in swine, but not confirmed
  - as of mid-2009, only two documented cases of swine-origin H1N1 influenza A virus in pigs worldwide

9.3 Give four characteristics of viral replication that support Darwin's theory of natural selection? Illustrate your answer with examples.

Viral replication leads to the production of large numbers of progeny

- poliovirus produces approximately 10^5 virus particles per infected cell
- 10^9 HIV particles are produced per day in an HIV infected person
- 10^{11} hepatitis B virus particles are produced per day in a hepatitis B virus infected person
- Darwin stated that organisms produce more progeny than necessary to ensure the organism's survival

Viral replication introduces mutations into the genomes of the progeny

- RNA viruses have mutation rates of 10^{-4}
- viral RNA polymerases have no proofreading ability
  - for a genome of 10,000 nucleotides, one mutation is introduced per genome replicated
- population of RNA viruses not homogenous
  - termed "quasispecies"
- medically important in, for example, influenza A virus
  - errors in viral replication lead to appearance of new strains every 2 to 8 years
  - if new strains not recognised by immune system, influenza outbreaks may occur
  - termed "antigenic drift"
- DNA viruses have much lower error-rates
  - between 10^{-7} and 10^{-8}
  - but still higher than those of cellular systems
- Darwin stated that offspring always differ from the parents

Viral numbers are reduced by host immune system, by exposure to heat, by drying and by competition for a cell or organism

- HIV particles synthesised in infected host mostly destroyed by immune system
  - virus almost undetectable in blood
- hepatitis B virus infected hepatocytes destroyed by cytotoxic T-cells
  - liver damage caused by host defence, not by cytolytic activity of the virus
- competition for hosts seen in the wild
  - new foot-and-mouth disease virus type O strain first seen in India in 1990
  - gradually displaced all circulating foot-and-mouth disease virus strains in Asia
  - spread to United Kingdom in 2001 to cause major outbreak
- Darwin stated that numbers of an organism are reduced by environmental conditions, competition for food and for habitat.

Certain viruses can be adapted in the laboratory to replicate in different cell lines or organisms

- mutations occurring through inaccurate replication give advantage in the new habitat
- important in the production of live viral vaccines
  - e.g adaptation to replication in monkey cells of poliovirus leads to loss of virulence in humans
- Darwin observed that variation produced in offspring could be used by farmers to generate new breeds of livestock.

Suggested further reading

Textbooks

- Principles of molecular virology. Alan J. Cann (Academic Press).
- How pathogenic viruses work. Lauren Sompayrac (Jones & Bartlett Publishers).
- How the immune system works. Lauren Sompayrac (Blackwell Publishers).
- Introduction to modern virology. N. J. Dimmock, A. J. Easton & K. N. Leppard (Blackwell Publishing).
- Lecture notes on infectious diseases. B. G. Mandal, E. G. L. Wilkins, E. M. Dunbar, R. T. Mayon-White (Blackwell Publishing).
- Principles of virology – molecular biology, pathogenesis, and control. S. J. Flint, L. W. Enquist, V. R. Racaniello, A. M. Skalka (ASM Press).

General reading

- Catching cold. Pete Davies (Michael Joseph).
- Guns, germs and steel. Jared Diamond (W. W. Norton & Company).
- The hunt for a killer virus: Hepatitis B. Baruch S. Blumberg (Princeton University Press).
- Witness to AIDS. Edwin Cameron (Tafelberg Publishers).

Index